Musician's Acoustics

Scott E. Parker, Ph.D.
The University of Colorado
Department of Physics

Jamison A. Smith, Ph.D.
The University of Colorado
Laboratory for Atmospheric and Space Physics

Library of Congress Cataloging-in-Publication Data

Parker, Scott E. and Jamison A. Smith
Musician's Acoustics / Scott E. Parker and Jamison A. Smith
Includes bibliographical references

Printed in the United States of America

ISBN-13: 978-1482566338
ISBN-10: 1482566338

Cover image courtesy of Finke Horns: www.finkehorns.de

Preface

This book originated from class notes for two musical acoustics courses I have taught at the University of Colorado in Boulder for over 15 years. The target audience is musicians, including students, and somewhat serious part-timers with day jobs. The science of sound involves some very complex concepts that this book will not shy away from, but the math and physics will be kept to a minimum and can be skipped without losing too much qualitative understanding of the material.

I am a full-time physicist who plays jazz saxophone for fun. My research in plasma turbulence gives me a solid understanding of the underlying physics of sound waves, but it has been through teaching musical acoustics courses that I have built the knowledge base for this book. The field of musical acoustics is an area where art, physics, and psychology overlap, producing striking and sometimes surprising results. The primary goal of this book is to provide an introduction to the science of sound that is useful to the active musician. This book could be used as a supplemental book for a semester survey course for students majoring in fields outside of the sciences. However, it is not a textbook, in the sense that it does not have the usual exercise problems at the end of each chapter. In our course, at the University of Colorado, homework is now done on-line, and concept questions are asked real time in class.

One challenge in presenting the topic of musical acoustics is that many of the physical concepts are quite deep. This book, however, focuses on music and uses physics for music's sake – not physics for physics' sake. I bend over backwards to avoid in-depth physics discussions that might lead the reader to ask, "What does this have to do with music?"

Anyone involved in creating electronic sound or producing recordings will want some understanding of the concepts covered here, as musical acoustics is the starting point for understanding reverberation, compression, vocal presence, equalizers, data compression, sampling, and synthesis. This is not a technical "how-to" manual, but rather a "how-things-work" book. It is our hope, that you, the musician, can further explore the ideas presented in this book in your everyday practice and performance.

The authors thank the numerous people who have provided figures in this book. We also thank Helen Parker and Gary Benson for help with editing the text.

- SEP, Boulder, CO 2013

Table of Contents

Chapter 1 Vibrations and Sound

Chapter 1 Outline

- Vibrations
- The Simple Harmonic Motion
- Resonance
- Audible Frequency Range of Hearing
- Complex Periodic Waveforms
- Pressure
- Sound
- Speed of Sound

In this book, we talk about musical sound. We will talk about a lot of stuff that is important to the musician, but this book does not contain the "magic recipes" with which you can immediately run into the studio and record an awesome new track. This book contains background information about your musical source – horn, "axe," voice, or whatever you play – and your musical appreciation system, mostly your ears and brain. This foundation of knowledge is called musical acoustics, and our hope is that you will build your own musical foundation from the material contained here and use that foundation to improve your craft.

You may be mildly disappointed that we are not going to give you the secret to create better music.... Maybe you already know there is no such secret. Music has mathematical and physical rules, but music is also an art. Its beauty depends on both the skill of the performers and the expectations of the listeners.

We are going to start out with something musically, which is so very simple, but yet clearly one of the most fundamental aspects of music: a steady tone. Think of one note – say a whole note... just one note, not a composition, no harmony, no rhythm... just... one... note.

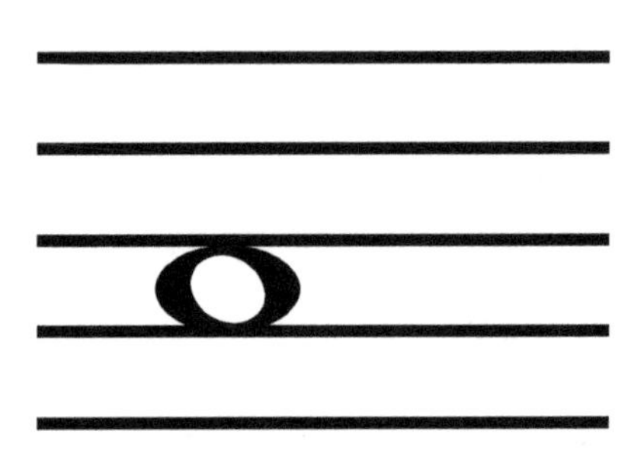

Vibrations

The steady tone is the fundamental building block of music, and we want to understand it well. As musicians, we know that "the beat" or rhythm is also just as important, but we'll focus on the tone for the time being. Regardless of the instrument, we care about the unique sound we produce. We also care about variations on the steady tone: attack, decay, vibrato, staccato, falsetto, pedal tones, subtones, harmonics, and more. In order to get a steady tone (a note), we need "something" to vibrate. For example, when you strike a bell, what does it do?

Well, it rings obviously, but what does this mean? The ringing is a manifestation of the bell **vibrating** (you can feel it if it is a big low pitched bell). The vibrations also decay away. Not all bells produce a distinct tone. Some bells do, some bells don't.

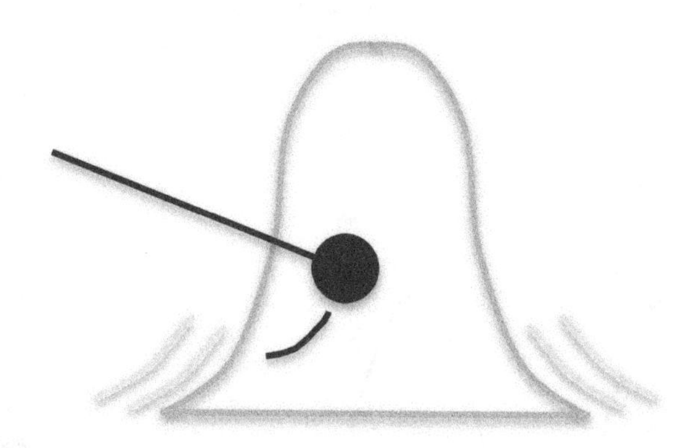

This vibrating motion is ubiquitous in the physical universe. Not only do a variety of musical instruments (bells, vibraphone bars, strings, symbols, gongs) vibrate, but so does the Sun and the Earth. The energy released by a strong earthquake rings the Earth like a bell – at extremely low frequencies. Earthquakes also excite natural modes of vibration in structures, such as buildings and bridges, that can cause the structures to crack and collapse – a big problem for architects and civil engineers. The Sun vibrates at particular frequencies, as do atoms and the electrons within an atom. Much of quantum mechanics is calculating the frequency and shape of particle vibrations, such as electrons within an atom or quarks within a proton.

Quantum mechanics, while interesting, is irrelevant to sound waves, and this book is about the science of sound. What is sound? Why do things sound the way they do? How does sound get from one place to another? The answers to these kinds of questions are what we are after in this book.

Air is actually quite springy. You can compress it and stretch it. Think about a balloon or an empty two-liter plastic Coke bottle with the top tightly secured. You squeeze it, but then it will spring back when you release it (it will make that crackly sound as well).

Simple Harmonic Motion

To get an idea of what a wave or vibration "looks" like, let's talk about springs first. Let's take a look at a mass on a spring, the simplest of all vibrating objects.

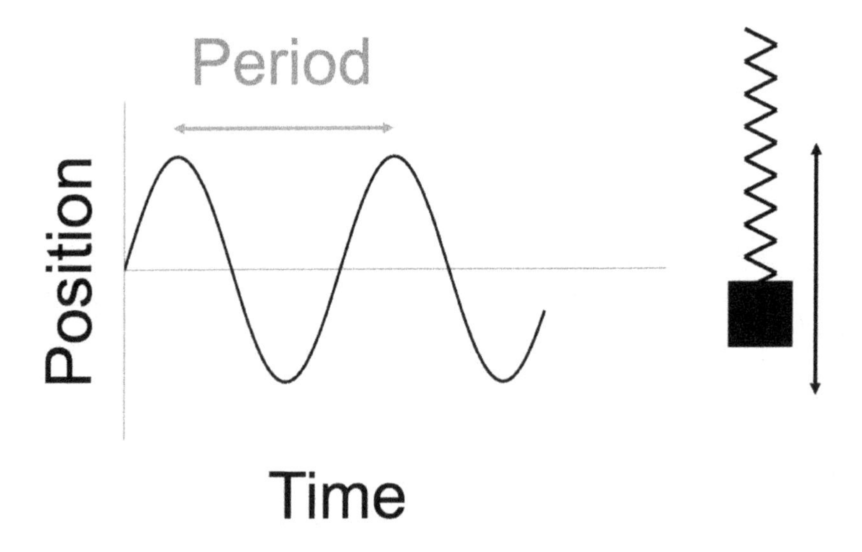

A mass on a spring is an example of a **simple harmonic oscillator**. If we plot its position over time, we produce a sine wave. Its up and down motion repeats itself at regular intervals. This property of repeating at regular intervals is called periodicity – or we say the motion is **periodic**. The time it takes to repeat one cycle of the periodic motion is called the **period**. For a typical spring, the period might be a fraction of a second long. Sounds like a lot of jargon, doesn't it? Well, we need to precisely define these things (and a few more) to really understand steady tones and pitch.

The period of a simple harmonic oscillator is related to the oscillator's **frequency**.

$$period = \frac{1}{frequency}$$

Two things control the frequency of a simple harmonic oscillator: the stiffness of the spring and the mass. The actual equation that describes the frequency of a simple harmonic oscillator is the following:

$$frequency = \frac{1}{2\pi}\sqrt{\frac{stiffness}{mass}}$$

A stiffer spring will have a higher frequency than a less stiff spring. Also, an oscillator with more mass will have a lower frequency than an oscillator with less mass. As stiffness increases, the frequency increases:

stiffness ⇑ frequency ⇑

As stiffness decreases, the frequency decreases:

stiffness ⇓ frequency ⇓

There's an inverse relationship between mass and frequency. Heavier objects vibrate with a lower frequency and vice versa:

mass ⇑ frequency ⇓
mass ⇓ frequency ⇑

Resonance

If we give the mass on the spring very slight pushes at regular intervals equal to the period of oscillation, the amplitude will increase over time. The simple harmonic oscillator will gain a little energy with each swing through its cycle. This is known as the concept of **resonance** – driving simple harmonic motion at its natural frequency. A key aspect of resonance is that the frequency of the oscillator is independent of amplitude. This is not usually the case. For example, a pendulum's frequency depends on the size of its swing. You may think of the pendulum device that is found in one of those large old-fashioned grandfather clocks. For the pendulum to keep accurate time, the distance that the pendulum swings must be kept small. It turns out that many vibrating objects behave as simple harmonic oscillators at small amplitude vibrations.

Audible Frequency Range of Hearing

If an object vibrates with periodic motion and with a frequency in the audible range, it will produce sound that has the sensation of a distinct musical pitch. The **pitch** is the perception of a musical note. The pitch is often described as being "high" or "low." The **audible range** for humans is roughly 20 Hertz to 20,000 Hertz.

Hertz is a measure of frequency – abbreviated as Hz. A frequency of 1 Hz corresponds to a period of 1 cycle of motion every second. A frequency of 20 Hz corresponds to an oscillator vibrating through 20 cycles every second – much faster than most springs and much faster than you can wave your hand up and down.

It is important to know about the 20-20,000 Hz audible range to understand what objects might produce sounds and what frequencies and time periods we need to consider when manipulating digital sound. The actual frequency range of hearing is a little narrower than 20-20,000 Hz and also depends on the age and gender of the listener. We will say more about the audible frequency range of hearing later in Chapter 6.

Pressure

Sound waves are very small (but fast) deviations in the background air pressure, so we need to cover just a little about what pressure is at this point. We will cover pressure in more depth later as well. **Pressure** is the force per unit area. **Force** is measured in Newtons (N) and it is the amount one object "pushes" on another object. One **Newton** is about one-quarter of a pound. We are immersed in a background air pressure of approximately 100,000 N/m^2! This is actually a lot of pressure and is the result of the weight of the atmosphere. Sound waves are small fluctuations in the background air pressure in the range of 1 N/m^2 or less. The fact that atmospheric pressure is so huge and that sound waves are such small changes in the background pressure is quite remarkable. We will have more to say about sound waves in Chapter 3.

Sound

Let's take a look at this speaker connected to something called a signal generator (it generates an oscillating signal that we will talk about more soon). It could, as well, be a speaker hooked to any source (e.g. a computer) producing a steady tone. The signal is a fluctuating voltage that drives the flow of electricity within the speaker.

The varying flow of electricity within the speaker causes the cone to vibrate and thus become a source of sound. The cone pushes forward on the air creating a region of high pressure, called a **compression**, and then the cone pulls backwards, creating a region of low pressure, called a **rarefaction**. These compressions and rarefactions travel as a wave to your ear and cause your eardrum to vibrate. The pressure variations are very small – typically a million times smaller than the background air pressure. These tiny variations in air pressure are what create the sound for your ears to hear.

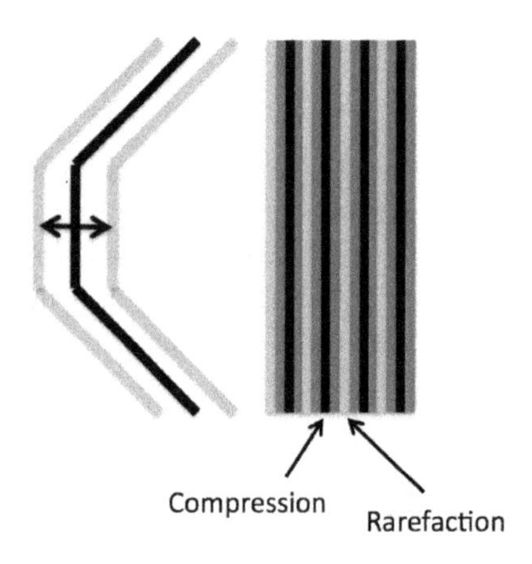

Complex Periodic Motion

Not all periodic motion is as simple as the simple harmonic oscillator. Many objects vibrate with several natural modes of vibration simultaneously (as if attached to several harmonic oscillators simultaneously). The vibration of most objects can be described as a collection of simple harmonic oscillators. For example, if you strike the fifth string of a guitar, that string will vibrate with several natural modes of vibration. Each of these natural modes has its own frequency. The lowest frequency produced by one of these modes is 110 Hz. Some of the other modes have frequencies of 220, 330, 440, and 550 Hz. Together, this mixture of vibrational modes produces a **complex periodic waveform**. Take a look at this complex periodic waveform produced by the bowing of a violin string.

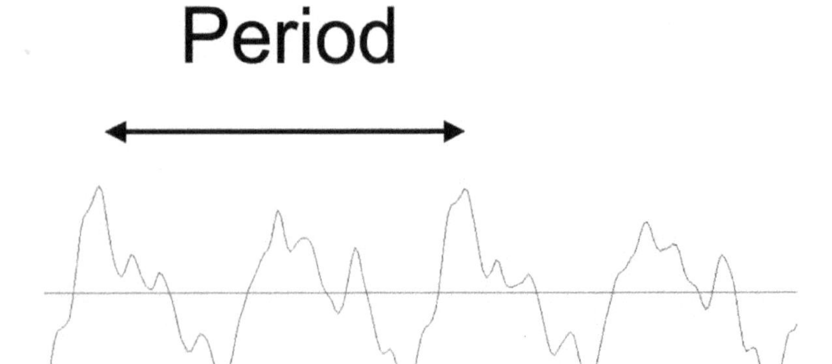

The precise mixture of contributions from the various natural modes is part of the nature of musical sound that we call **timbre**. Timbre allows our ears to distinguish among the various playing instruments. We can tell if we are listening to a trumpet or a piano – even if those two instruments are playing the same note with the same period for their individual complex waveforms. The mixtures of frequencies – combined with the other aspects of sound, like attack and decay – allow us to distinguish the different instruments contributing to the same note.

Now, there are other musical instruments that do not produce the sensation of pitch (mostly percussion instruments... drums, cymbals and gongs), yet they are still musical. Why are many of the percussion instruments different? Here is the complex waveform from a crash cymbal. For the ear, it produces a NOISE, rather than a musical pitch. Notice how there's no period; the oscillating motion does not repeat itself at regular intervals. The ear interprets this aperiodic vibration as either a strange sound (e.g. like a gong sound) or as noise. For an aperiodic vibration, there is no distinct sensation of pitch.

Speed of Sound

The speed of sound is approximately 344 meters per second (or 344 m/s) in air under normal conditions. You may have been in a few situations where you experience a time delay in the arrival of a sound. For example, at a baseball park you might see the baseball hit the bat, yet experience hearing the sound (a "crack") a fraction of a second later. The speed of light is very fast (3×10^8 m/s). As far as sound is concerned we can assume that we see events happening instantaneously.

The speed of sound in air depends on temperature. Sound travels a little slower on cold days and faster on warm days. The speed of sound (we will call it v_s in this book) is given by

$$v_s = 344 \text{ m/s} + 0.6 \times (T - 20) \text{ m/s}$$

where T is the temperature in Celsius.

Example:

Suppose that you saw Todd Helton hit a baseball in Coors Stadium. How long would it take to hear the sound 50 m away when the temperature is 30°C?

Solution:

$$v_s = 344 \text{ m/s} + 0.6 \times (30 - 20) \text{ m/s} = 350 \text{ m/s}$$

$$t = d / v_s = 50 \text{ m} / (350 \text{ m/s}) = 0.14 \text{ s}$$

We covered a lot in this chapter, rather quickly. Throughout the book we will re-investigate these concepts and they will become much clearer. We need to fill out more details, show more examples, and get comfortable using these new ideas. In summary, *vibrating motion that is periodic* (periodic means repeats itself at regular intervals) *produces the sensation of a distinct pitch*. The simplest periodic motion is produced by the so-called simple harmonic oscillator which is a conceptual idealization of the simplest type of oscillator (mass-spring). Things that do not exhibit periodic motion do not produce a distinct pitch.

References and Resources

At the end of each chapter, we will give a short list of references for further reading. Here, for Chapter 1, we simply list three well-known introductory textbooks.

Benade, A., Horns, Strings and Harmony, 1960, re-published in 1992– *A classic book on musical acoustics*

Hall, D., Musical Acoustics, 3rd Edition, 2002 – *excellent non-science textbook* on the topic

Rossing, T., F. Moore, and P. Wheeler, The Science of Sound, 3rd Edition, 2002 – *an excellent college textbook covering the whole field of musical acoustics*

Chapter 2
Digital Sound

Chapter 2 Outline

- Digital Sampling
- Microphones
- Sound Cards
- Digital Sound Editing Software
- Sampling Properties

The use of computers to record and manipulate sound has revolutionized our day-to-day use of recorded music. What was done 20 years ago in expensive recording studios can now be done with an everyday notebook computer.

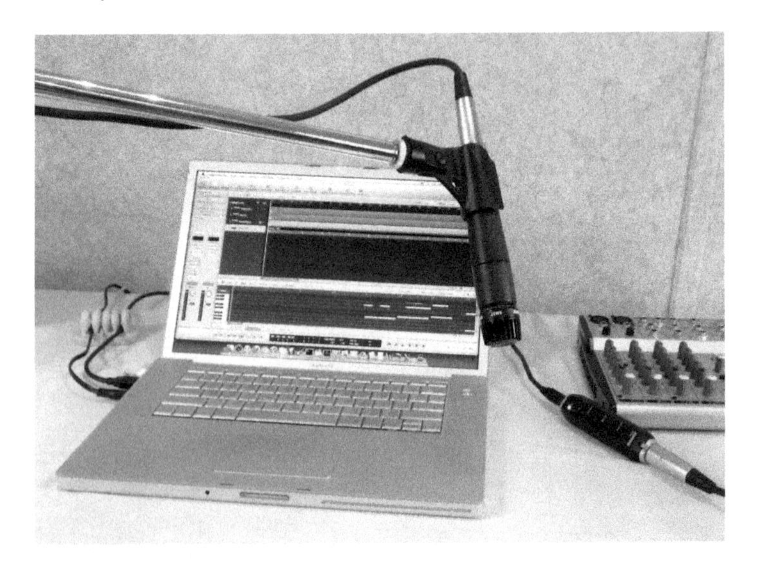

The sound waves emanating from our voice and musical instruments are analog waveforms, but the computer records a digital representation of that original analog waveform. Not only does digital sound impact practical aspects of today's music, it can help us tremendously understand and interpret musical sound. So often when we discuss waveforms, we are talking about both an abstract physical concept and an actual digital sound sample on a computer. In this chapter, we will further develop our understanding of these waveforms and how they are stored and manipulated on the computer.

In the last chapter, we discussed the concept of a complex periodic waveform and how it was special in that it produced the psychological sensation of distinct pitch. Often, we will drop both adjectives "complex," meaning that it is not sinusoidal and "periodic," meaning that the waveform repeats itself at regular intervals. We simply will call the digital sound a wave or waveform. These terms are used interchangeably in digital sound: waves, waveforms, samples, and wave files.

Digital Sampling

The best way to really understand complex periodic waveforms is to see one and manipulate it using a digital sound editor. A good, freely available digital sound editor is called Audacity®. All digital sound software does a similar thing when it records and stores musical sound. Different software packages, however, do many more things than simply record, store, and edit sounds. Each package fills a different niche in the musical industry. For example, Audacity®, Sony's Sound Forge™, and Adobe® Audition® are simpler programs and better for manipulating raw individual sounds. Software packages like Pro Tools®, Apple's Logic Pro, and Steinberg's Cubase are better for mixing (like done in a recording studio) and musical production. The Raven software program from the Cornell Laboratory of Ornithology at Cornell University is specialized for bioacoustics research.

We recommend one of the simpler software tools, probably Audacity® or Raven Lite as an introduction to digital sound because these two programs are both simple to use and free. There are many, many digital sound software packages, and you can no doubt name a few that we have not mentioned. Let's simply go through the process of recording a digital sample starting from the microphone, the sound card, and finally the digital sound editing software that stores the waveform and manipulates it.

Microphones

There are a variety of ways to convert sound waves into an electrical signal. Common types of microphones include dynamic, condenser, electret condenser, and ribbon microphones.

In addition to labeling a microphone by the way it converts sound waves into an electric signal, microphones are described by the spatial pattern of their sensitivity, such as omnidirectional, cardioid, and shotgun. Omnidirectional mics pick up sounds in all directions. Cardioid mics are designed to be sensitive for sounds coming "head on" directly to the microphone. Shotgun mics are designed to be very, very directional – as in super-cardioid.

Sometimes, one describes a microphone by make and model. For example, "It's a Shure SM58." A microphone may be referred to by its directionality such as, "It's a shotgun mic," or its transducer as in, "That's an electret condenser."

Condenser mics involve the motion of a diaphragm changing the capacitance of an electrical circuit and thus require an external power source. Condenser mics are very nice, very sensitive, and produce a warm sound. A good condenser mic can be very expensive. For example, a Neumann U87 large diaphragm condenser microphone can cost a couple thousand dollars. There are cheaper condenser mics as well that work in some situations. Condenser mics tend to be very sensitive and pick up less desirable sound qualities of the room. Care is needed to get a good sound from a condenser mic. The Neumann condenser mic shown below sells for around $4,000.

The Neumann U 87 Condenser Microphone, Courtesy of Georg Neumann GmbH.

For professional recording and sound re-enforcement, cables with XLR connectors are used, and power is supplied to a condenser microphone by the mixer through the XLR cable at 48 Volts DC. The power used to drive condenser microphones is referred to as "phantom power."

Female (left) and Male (right) XLR connectors

Dynamic microphones are a little simpler and do not require phantom power. They produce a fairly strong signal without the need of an external power supply. Dynamic mics are also nice because they are solid – often pretty indestructible – and usually pretty inexpensive. The industry standard is the Shure SM58 for voice and Shure SM57 for instrument sound re-enforcement. You will see these microphones regularly in sound re-enforced venues (for example, a lecture hall, a hotel ballroom, a band stage). A SM58 or SM57 sells for around $100. There are less expensive microphones that do work well. Keep in mind though, for a public speaking event or musical performance, few are willing to take chances and would rather choose the quite reliable SM58.

The Shure SM57 and SM58. The SM58 (top) is a commonly used dynamic microphone for vocals. Images © 2013, Shure Incorporated - used with permission.

Let's discuss how a dynamic microphone works. First, sound waves pass through the head of the microphone that provides protection of the quite sensitive diaphragm. Additionally, the head structure can be acoustically quite complex and can change the directional sensitivity of the microphone and affect its overall acoustical response. Sound waves cause the pressure to increase and decrease at the mic diaphragm in the audible range (20 - 20,000 Hz). These pressure fluctuations push and pull on the diaphragm causing it to vibrate. The diaphragm is connected to a coil of wire, which moves in an external magnetic field supplied by a permanent magnet.

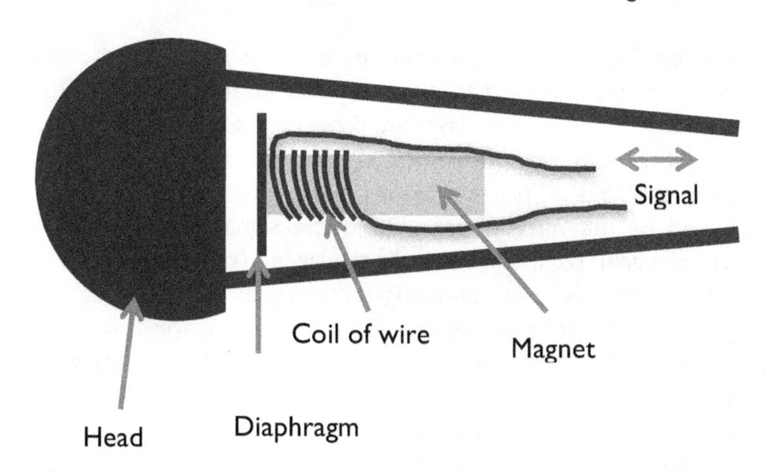

There is a law of physics called Faraday's Law that says that if the magnetic field passing through a loop of wire changes, it induces a voltage around that loop. In the case of the dynamic microphone, as the coil moves, the magnetic field passing through the coil changes. This, in turn, induces a changing voltage (an electric generator works by using the same principle). The microphone converts a pressure fluctuation into a varying voltage carried by the wires.

It is important to realize that the signal intensity of a microphone increases enormously as you decrease the distance from the sound source and the microphone itself. If you have had any experience singing with a microphone, you immediately recognize the importance of the location of the microphone and how you can control dynamics enormously by simply moving the microphone a little. Additionally, when singing near the microphone a vocalist can generate an effect called "vocal presence" where the low to mid-range frequencies are boosted. This can give a female vocalist a warmer sound. Similar effects can be added later by adjusting the EQ, or equalization, which means adjusting the intensities of different frequency ranges. Very subtle differences of directionality and vocal presence can make vocalists very picky about microphones. This is very similar to the pickiness of saxophonists and how they regard mouthpieces and reeds!

Sound Cards

In the simplest setup, the electrical signal from the microphone enters the computer's sound card by way of the cable connecting the microphone to the computer. There are many alternatives to using a sound card, but a similar process takes place regardless of the interface used. The term "sound card" dates back to the days when desktop PCs had a physical card plugged into a slot in the computer motherboard.

The sound card has a "preamp" that boosts (amplifies) the time varying signal produced by the microphone. Next, the signal is converted from a smooth analog signal to a discrete digital signal (represented as integer numbers). You can see this sampling process by using Audacity®, or another digital sound editor, to zoom in on a short time window.

The sound card takes discrete samples of the signal at a very small time interval, say a fraction of a millisecond. We'll quantify the details and give actual numbers in the next section. The electrical component that converts the analog voltage to a digital voltage (or integer) is called an analog-to-digital converter. The whole process is called **digital sampling**.

The figure below shows how digital sampling or analog-to-digital conversion works. A smooth, continuous signal is "sampled" at regular time intervals to produce particular values at discrete times.

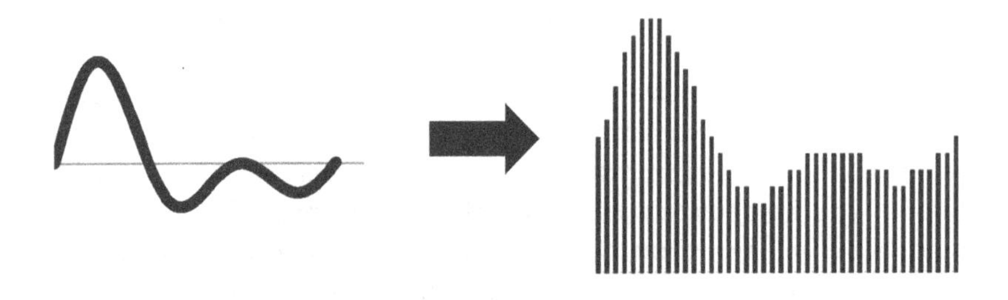

The sound card does the reverse process, digital-to-analog conversion, when your computer speakers play back the recorded sound. The digital sample is converted into an analog – meaning continuous – signal versus time. The analog signal can be made smooth by applying an appropriate electronic circuit that filters out high frequencies.

There are better choices to improve sound recording quality than simply plugging a high-gain, dynamic microphone into your computer's sound card. (That is, plugging the mic directly into the computer.) Typically, sound cards have a **bit depth** of 16 and a 44,100 Hz **sampling rate** which is CD quality (we will discuss these numbers in more detail below). Sound card quality, however, varies greatly. Very often, the gain coming from a better microphone is too low for some sound cards. A simple solution is to use a mixer. A mixer allows mixing multiple inputs (say microphones) into one stereo output. A mixer typically has a preamp for each input channel and the capability to adjust the EQ of each input channel individually. Also, most mixers (even small inexpensive ones with a few channels) have phantom power, which is needed for many mics (usually, condenser mics). One can then take the output from the mixer and send it to the sound card's "line in" jack. On most Macs, the "line in" and the "mic in" is the same 1/8" input. The choice is made by the operating system settings. On Windows PCs, the line in and mic in are usually two distinct 1/8" stereo inputs.

Another simple and very common solution is to use a USB audio interface. These devices can be thought of as a USB sound card in a box. Some of them are very good and well worth having. An excellent choice for laptop stereo recording would be a two channel USB audio interface that has a 24 bit depth and a 96 kHz sampling rate. These can be purchased for about $200. If we were buying one, we'd look for 24 bit, 96 kHz resolution and at least two XLR inputs with phantom power. You will find many, many devices that do not have these important capabilities. Most audio interfaces are the CD quality 16 bit, 44 kHz resolution range. Many do not have either XLR input or phantom power. For quality recording, it is good to have the higher resolution. This will be explained more in the next section.

There are yet simpler options that can work quite well for many applications, e.g. podcasting or home music production. First, you can find fairly good quality microphones with USB out. These microphones have the pre-amplifier built in and take power from the USB bus. You can, in principle, hook two USB microphones for stereo recording, but the computer system settings can get tricky. There are also mono USB audio interfaces that take the form of a plug or adaptor or cable. These are typically 16 bit devices, but very handy because they reduce cables, power supplies and boxes, if you want to do simple good quality recording of your musical instrument and still keep your desk clean or be portable. The quality of these interfaces is certainly high enough to develop a demo CD.

Digital Sound Editing Software

The digital sound editing software records the analog signal in a digital format, stores it, displays it, and plays it back. We have already briefly mentioned some of these software programs. The amplitude, frequency, sampling rate, and timbre of the waveform can be manipulated using various options and features. The sound can even be edited point-by-point at individual sample times. Here, we do not give justice to the digital sound-editing software. These programs can really do quite remarkable things, both by creating and modifying waveforms. Here, we are simply describing the mechanics of how the sound is converted from mechanical motion of air to digital data.

Sample Properties

Computers use binary integer numbers to represent data. So, we need to know a little (very little) about the conventions used when digital samples are stored on the computer. As an example, let's take the standard case, a stereo CD quality recording. CD quality is considered high quality, at least for playback. For CD quality, there are 44,100 samples taken each second. The abbreviation of second is "s."

CD quality sampling rate = 44,100 samples / s

In other words, one sample of the signal is sampled every 0.0000227 seconds. This is what is called the **sampling rate**. Sampling rate is usually indicated by its frequency in kiloHertz (kHz). 1 kiloHertz is the same as 1000 Hz. CD quality sound has a sampling rate of 44.1 kHz.

Two signals are stored, one for the left channel and another for the right channel. Each sample is represented as a 16 bit binary number. We call this the **bit depth**, and we describe the bit depth simply by saying "a 16 bit sample." Very often, you will see specifications of digital equipment stated as, for example, "24 bit 96 kHz" resolution where it is implied that the bit depth is 24, and the sampling rate is 96 kHz.

The bit depth is the number of binary bits used to represent the digital sound. This means – for a bit depth of 16 – using base-ten (the number system we normally use in everyday life), the signal runs from:

0 to 2^{16}-1
...or from 0 *to* 65,535
...or really from -32,768 *to* +32,767

We say "or really from -32,768 to +32,767" because the oscillatory signal goes up and down about zero. This is a pretty good quality sample, and it will sound good.

We can calculate how much data is involved in storing one second of CD quality sound.

1 s x 44,100 samples / s x 16 bits / sample
x 2 (for the two channels) = 1,411,200 bits

There are 8 bits within 1 byte, so:

1,411,200 bits / 8 = 176,400 bytes

CDs hold about 74 minutes (min) or:

176,400 bytes / s x 60 s / min x 74 min = 783,000,000 bytes

1 million bytes is approximately 1 megabyte and is abbreviated as MB. The label on your blank CD probably says the CD has a capacity of "750 MB." There's always some ambiguity about MB because sometimes it means 1,000,000 bytes, and sometimes it means 2^{20} bytes (2^{20} = 1,048,576).

Typically, for music, we want a good quality sample (CD resolution or better). You don't always want a CD quality sample rate and resolution because it uses a lot of memory. The digital sound editor gives you the option of choosing 8, 16, 24 or 32-bit resolution and sampling rates ranging from 8 - 192 kHz. In the end, the hardware – meaning the sound card specifications – determine how high a sampling rate and bit depth is really possible.

Our experience is that CD quality is pretty good, and it is hard to hear the difference between 16 bit 44.1 kHz and a higher quality sample. The higher bit depths and sampling rates, however, are important during the recording, mixing, and mastering process. There is a general rule in audio recording that you want to start with the highest resolution sound you can and reduce its resolution to CD quality in the end. We follow this rule because sound loses quality as it is processed and reprocessed. This is especially true for analog sound. Each step in mixing and processing degrades the quality – much like photocopying a document and taking a photocopy of the photocopy. Try to limit the number of analog-to-digital and digital-to-analog conversions involved in the process. One could argue, once the sound is digitized, it's quality can be maintained from step to step as it is manipulated in the recording/mixing/mastering process, but some degradation is to be expected with each additional manipulation.

It is important to discus one last concept: the Nyquist frequency or the highest possible frequency a digital sample can capture. We define the **Nyquist frequency** as simply:

Nyquist frequency = sampling rate / 2

The Nyquist frequency is an estimate of the highest frequency that can be represented for a particular sampling rate. There can be no sounds recorded with frequencies higher than the Nyquist frequency. Sounds with frequencies close to the Nyquist frequency will be poorly sampled and thus distorted. So, the Nyquist frequency is an important frequency to be aware of when recording and manipulating digital sound. For CD quality samples, the Nyquist frequency is 22,050 Hz, which is just above the audible range. For an 8,000 Hz sample, the Nyquist frequency would be 4,000 Hz and this would be where the frequency spectrum would be cut off. 4,000 Hz is well within the audible range, so this is not a good quality sample rate although it might be good enough for some things, like transmitting a phone conversation.

Digital sound can be compressed and decompressed to reduce file storage size and digital streaming bit rate requirements. The "lossy" compression technique called MP3 and the file-sharing program called Napster, developed by Shawn Fanning caused a major revolution in the music industry beginning around 1999. Compression is a huge topic and is not covered in this book. There are many, many different types of compressed digital audio files, which you may already be aware of. There are two broad categories or types of compression: lossy compression and lossless compression.

Lossless compression does not, in principle, degrade the quality of the data at all. The strategy is to look for holes or long periods of time where nothing happens and remove or reduce data, then add the data back in when decompressing. Lossy compression actually reduces the quality of the data, but tries to retain the key sonic features of the original raw data set. MP3 compression is a lossy audio compression. It uses psychoacoustic concepts, namely masking, to reduce the size of the audio file. When a louder sound occurs with a frequency similar to a softer sound the human ear does not perceive the softer sound as well. Researchers at AT&T Bell Labs developed a "code" using psychoacoustics to take advantage of this fact to compress recorded audio signals into smaller files. A codec, or coder-decoder, is a computer algorithm to compress and decompress digital data.

Codecs are typically released as standards and may or may not involve encryption to restrict copying. We simply touch on data compression, because it is a current issue in digital sound and high technology, and you should know at least something about the basic idea.

References and Resources

Kirk, R. and A. Hunt, Digital Sound Processing for Music and Multimedia, 1999

Middleton, C., The Complete Guide to Digital Audio, 2003

Rossing, T., F. Moore, and P. Wheeler, The Science of Sound, 3rd Edition, 2002

Sokol, M., The Acoustic Musician's Guide to Sound Reinforcement and Live Recording, 1998

Chapter 3
Sound Waves

Chapter 3 Outline

- Sound Waves as Pressure Fluctuations
- Wave motion
- Superposition
- Reflection
- Sound Wave Propagation and Echoes
- Wave Interference and Beats

In this chapter, we are going to introduce the concept of sound waves. We will also study wave motion in general. Why? Because all waves behave more or less the same way. If we take a step back and say, "OK, sound is a wave, and waves behave like this," we will have a better perspective on musical sound. Richard Feynman (1918-1988), the famous physicist from CalTech, put it best when he said:

"You've seen one wave, you've seem them all." - Richard Feynman

Sound wave propagation is especially important for the understanding of later chapters, including the topics of brass and wind instruments, as well as architectural acoustics. Waves are important to scientists and engineers. An interesting point is that they are very well understood and relatively easy to model. Waves are also quite ubiquitous. Examples include: standing waves in musical instruments, vibrations of architectural structures, water waves, motion of a drum head, seismic vibrations in planets and stars, electron vibrations in atoms, vibrations of nuclei, vibrations of protons, quarks, string theory, etc.

Let's us begin our discussion of sound waves by reviewing a few concepts from Chapter 1. Air is compressible and springy. If you compress it, it springs back. Try clogging the hole of a bicycle pump or squeezing an empty plastic bottle with the cap on it. It is this springiness of air, which supports sound waves. If we compress air in one place and let it then spring back, it expands and compresses air in a region near to it. Regions of expansion are called rarefactions. These compressions and rarefactions travel away from their source at the speed of sound, which is roughly 344 meters per second (or 344 m/s) in air under normal conditions. The speed of sound in air depends on temperature, so sound travels more quickly when the air is hot.

$$v_s = 344 \text{ m/s} + 0.6 \times (T - 20) \text{ m/s}$$

where T is the temperature in Celsius and v_s is the speed of sound. We keep saying the speed of sound "in air" because sound travels at different speeds through different materials, like water and stone. We'll discuss these ideas later. From now on, the "speed of sound" refers to the speed in air; we'll specify if we want to discuss speeds in other materials.

Sound waves are a little bit hard to understand initially because we can't see them. One way to learn to understand sound waves is to use advanced mathematics to calculate the forces of parcels of air on one another as they compress and expand. Such an understanding requires the use of partial differential equations -- which are taught after calculus. Luckily, we can gain an intuitive feel for sound waves and their behavior without using advanced mathematics.

One curious property of sound waves is that the air molecules themselves are not moving from the sound source to your ears. The *sound wave* moves from the source to your ears at 344 m/s, but the actual air molecules don't blow from my mouth towards your ears at 344 m/s – that would be like getting hit with a baby tornado every time I uttered a word.

The air molecules *are actually moving* at a speed comparable to the speed of sound, but there are 10^{25} – ten trillion trillion – air molecules within every cubic meter. Those molecules are traveling in all directions, so they bump into each other quite often. An air molecule can only travel in one direction for a fraction of a second before it collides with another air molecule and changes direction.

It is not the air molecules that travel from the source to your ears. It is the disturbance in air pressure – the compressions and rarefactions – that travel from the source to your ears. This is true of all mechanical waves. The "oscillating stuff," called the **medium**, moves back and forth a short distance, but the wave – the disturbance – travels much larger distances, like ocean waves.

Sound Waves as Pressure Fluctuations

Sound waves are very small pressure fluctuations oscillating at frequencies in the 20-20,000 Hz audible range. **Pressure** is the amount of **force** pushing on a surface *divided* by the area. Force is measured in pounds or **Newtons** (N). One Newton is roughly equal to a quarter pound. A certain force exerted over a large area results in a lower pressure than that same force exerted over a smaller area. For example, if you are standing squarely on your two feet and then raise your weight onto the ball of one foot, you have just increased the pressure on the floor under the surface below the ball of your foot dramatically. You can feel the added pressure on the ball of your foot. A simpler experiment, but less dramatic, is if you lift one foot off the ground.... You exactly double the pressure on the surface below the foot still on the ground. (There is a subtle detail that you cannot really keep your balance without slightly adjusting the pressure at different locations on the bottom of the foot still on the ground.) In either example, the force that your body pushes down on the Earth is the same, but the area of contact has decreased. Likewise, you can decrease the pressure that you exert on a patch of thin ice by lying down on your stomach, thus distributing your weight over the large area of your torso. This is good advice for someone caught out on thin ice!

The atmosphere has a weight to it, and it pushes down upon you with a force of 15 pounds on every square inch of your body. This is actually a huge amount. That means there is over a ton of force pushing on your chest from the air directly above you – are you having trouble breathing? Probably not, your body is accustomed to this amount of pressure. An equilibrium exists between the pressure inside the inner cavities of your body and these external forces.

The background air (or atmospheric) pressure is approximately 100,000 N/m^2. We can express pressure in a variety of units. A Pascal (Pa) is the standard unit of pressure and is the same thing as a N/m^2 (1 Pa = 1 N/m^2). Pressure is also often measured (or expressed) in units of atmospheres (atm). For our purposes, we will approximate 1 atm as 100,000 N/m^2 (the exact value is 101,325 N/m^2) The pressure fluctuations that make up sound waves, however, are much smaller than atmospheric pressure. The pressure fluctuations associated with the sound from traffic noise are only about 0.1 N/m^2. That's actually pretty amazing. A rather loud noise is really only a pressure fluctuation that is 1 millionth of the atmospheric pressure! Yet, the ear hears the sound, no problem. In fact, the problem is that it is too loud a sound!

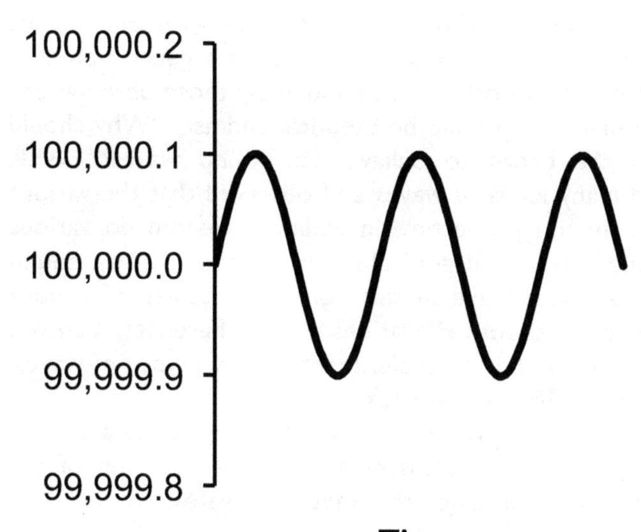

Zero pressure would be way down there

⇓

Wave Motion

A good way to understand sound waves is to examine wave phenomena in other media (the plural of medium). Wave motion is a generic physical phenomenon, so we can observe other waves and apply those observations to sound waves. Of course, you should be skeptical and ask, "Why should we expect waves on the ocean to behave like sound waves?" Well, physicists have studied many kinds of waves and observed that the various types of waves have common properties. In addition, we can do various experiments with sound waves, and gradually gain trust that, "Yes, sound waves do behave like a water wave in this context." Studies of various waves, however, have revealed both similarities AND differences. You will need to keep this fact in mind. For example, there are two types of waves: **transverse waves** and **longitudinal waves**.

Waves on a string, water waves, and electromagnetic waves are examples of transverse waves. For transverse waves, the motion of the medium is perpendicular to the direction the wave propagates.

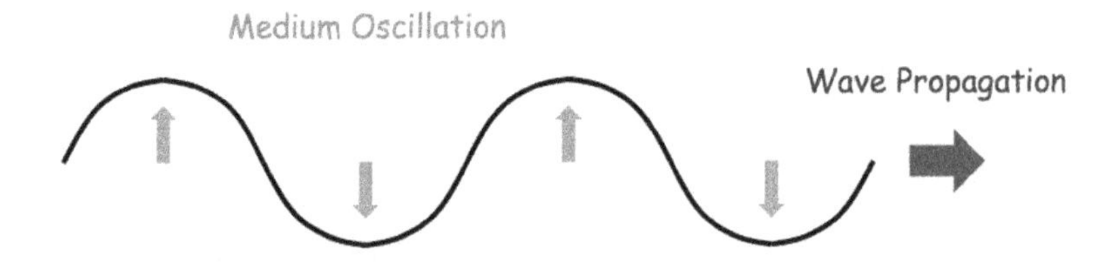

Sound waves and some Slinky waves are examples of longitudinal waves. For longitudinal waves, the motion of the medium is parallel to the direction of wave propagation.

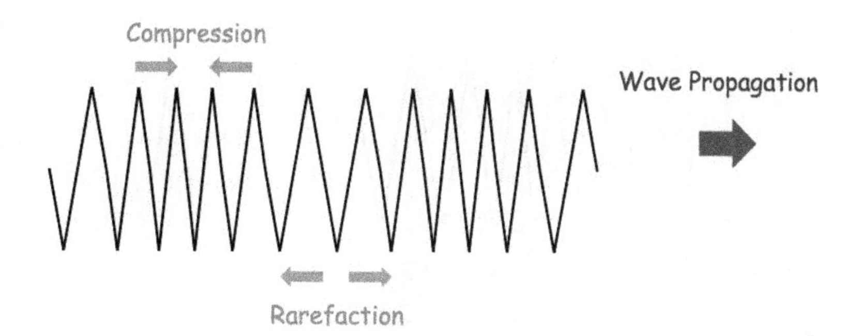

There are a few important properties of waves we need to understand. The distance it takes for a wave to repeat its "shape" is called the **wavelength**. The wavelength of a transverse wave is the distance between high points, also called **crests**. The distance between low points, called **troughs**, is also equal to the wavelength. For longitudinal waves, the wavelength can be measured as the distance between compressions or the distance between rarefactions.

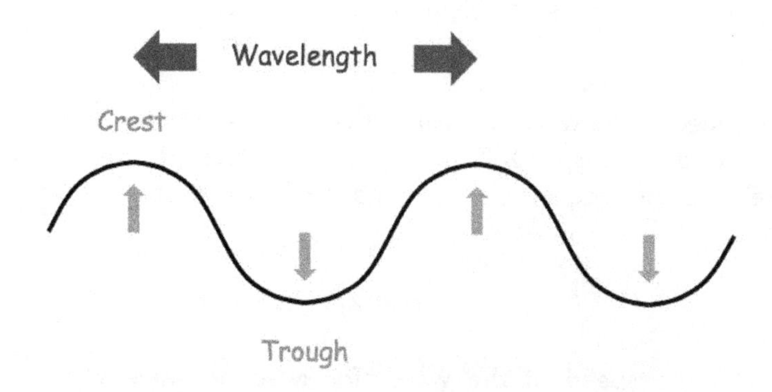

This is a transverse wave and the motion of the medium (the string) is up and down

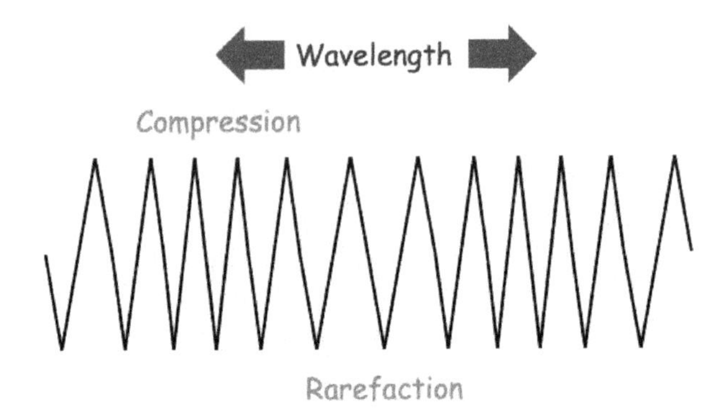

This is a longitudinal wave and the motion of the medium (the spring) is side to side

In Chapter 1, we defined the period to be the time it takes for a wave to repeat its shape. We also defined the frequency as the number of oscillations that take place every second. The period and frequency are related:

$$frequency = \frac{1}{period}$$

The speed of the wave is equal to the distance that the wave propagates every second. The speed is simply the product of the wavelength (a distance) and the frequency (the number of wavelengths that pass by you every second).

$$v = \lambda f$$

where v is the speed of the wave (for a sound wave $v=v_s$) , λ is the wavelength, and f is the frequency. λ is the lower-case Greek letter "lambda" and is commonly used to represent wavelength in all contexts.

Sound waves (under most situations) have a wave speed that is independent of the amplitude of the wave and the frequency. If the wave speed depends on amplitude, we call it a nonlinear wave. If the wave speed depends on frequency, then we have a dispersive wave. Nonlinearities and dispersion are important features in determining timbre. For example, the effect of finite string stiffness causes wave dispersion, and this is critical in determining the way a piano sounds. Dispersion is easy to handle mathematically. Nonlinearities are usually very hard to model (shock fronts, chaos, turbulence, etc.).

Superposition – Waves just add

Another important property of linear waves is the principle of linear superposition, which is simply the following: when two waves cross paths their displacements add together. This is shown in the figure. Two pulses travel towards each other, and three different times are shown. Then, two triangular waves are shown traveling through each other. It does not matter what the pulse shapes are or how big they are. They simply add together to get the total displacement at any instant in time.

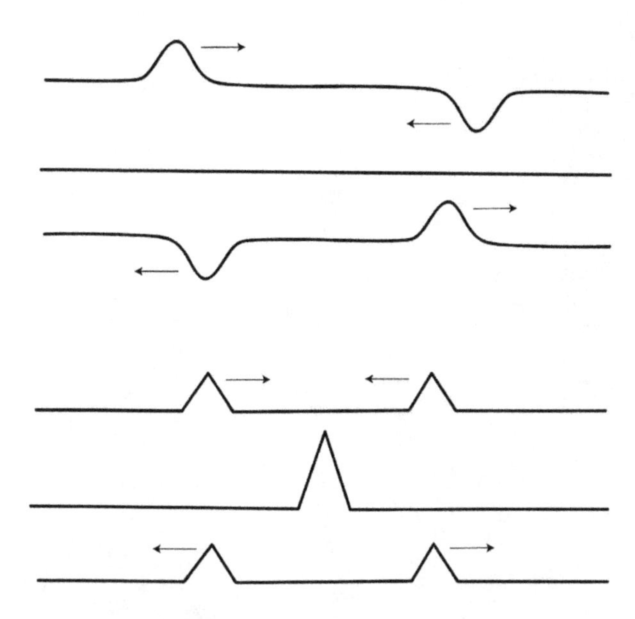

Waves **interfere** with each other in this very simple additive way. This is what causes the sensation of beats and other wave interference. **Beats** are periodic pulses in volume produced when two sound waves having similar frequencies interfere.

Another interference experiment produces "live" spaces and "dead" spaces. You can generate a sinusoidal sound wave at say 800 Hz using digital sound editing software such as Audacity® and connect the computer output to two external speakers. Next, if you move your head around a few feet to the left and right you will hear spatial maxima and minima in sound intensity (identical to beats, but in *space*, instead of time). Try it! This is not an experiment that you can do with headphones.

Very large amplitude waves, like sonic booms, do interact and do not simply pass through one another. This is another form of nonlinearity and would be an unusual circumstance in acoustics and unimportant in almost all musical situations.

Another reason we introduce the concept of superposition is that it helps explain another common and important phenomenon: **reflection**.

Reflection

In the figure above, we can have trouble actually doing this experiment because the wave pulse reflects at the end, and the reflection can confuse the issue. To do the experiment above, we launch the pulse, watch it for a short time, and ignore the reflection. Understanding reflection is crucial for understanding acoustics (concert halls, violins, or any instrument). The basic idea is that when the wave hits a rigid end (or fixed rigid wall in the case of sound or water waves), the pulse is reflected with opposite sign. If you go back to the experiment above with the 20 feet of rope tied at one end, we can see this behavior if we watch the rope carefully. See the following figure.

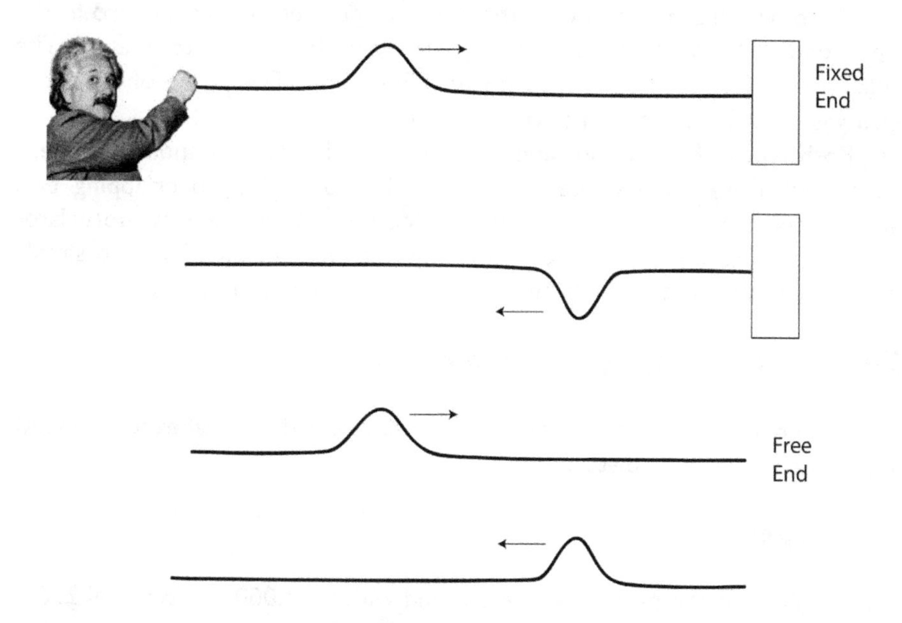

The reflection at the fixed end makes sense because in order for the medium (the string in this case) not to move at the fixed end, an equal and opposite pulse must be reflected so that the two pulses, the incident wave and the reflected wave, exactly cancel at the location of the fixed end. Amazing really! This might seem a little magical, but there is really no other way for the fixed endpoint to stay fixed unless there is a pulse moving in the opposite direction with opposite phase (exactly the same as the incoming wave except negative).

There is another type of reflection, which is not quite as intuitive, but is important when we discuss sound waves in wind instruments, e.g. clarinets and flutes. This is shown in figure above with the "free end" and is what would happen if you made a light whip-like motion with a whip or string not attached to anything (or attached to a slider, allowing up-and-down motion only). When the pulse reaches the very end, the displacement simply goes up, then down, and generates a pulse going the opposite direction (as shown). You can try this by whipping a rope or hose. This is a little tricky to demonstrate.

Suppose we had a pressure pulse traveling down a tube closed at one end. The refection is similar to the refection of a rope wave from a free end. It would slam into the wall (or closed end), then return traveling in the opposite direction with no phase change.

If the pipe has an open end, the pressure *fluctuation* goes to zero at the open end. (The pressure does not go to zero. It is equal to 1 atm.) The reflection is like the rope wave off of a fixed end. The phase of the wave changes by 180°, meaning the wave flips over.

Both types of reflections and are important for understanding what are called **standing waves**. Standing waves are formed by overlapping two waves traveling in opposite directions. We will talk about this more later with respect to standing waves within musical instruments. We are simply mentioning it now to emphasize the importance of wave reflection.

Sound Wave Propagation and Echoes

We now know quite a bit and can calculate the time it takes for a sound wave to travel a given distance.

Example:

What is the time delay for a sound with f = 1,000 Hz to travel 200 m at 30°C?

$v_s = 344 \text{ m/s} + 0.6\ (30 - 20) \text{ m/s} = 350 \text{ m/s}$

$t = d/v_s = (\ 200 \text{ m}\ )\ /\ (350 \text{ m/s}) = 0.57 \text{ s}$

Note that f = 1,000 Hz is not needed to solve the problem since the speed of sound is independent of frequency.

Wave reflections turn out to be very important in architectural acoustics. We will come back to this point in Chapter 12. Indoors, we are affected by sound reflected off walls all the time. But, while in a concert hall, we don't say or think, "Gee whiz, those are really nice sounding wave reflections." Rather, we simply sense the quality of whether the hall has "good" acoustics or not. Let us calculate the delay time for an echo to make sure we understand the distance the sound travels in such situations. Remember, in the situation of an echo, the sound has to travel to the wall and back, so the distance the sound travels is twice the distance to the wall. See the following figure.

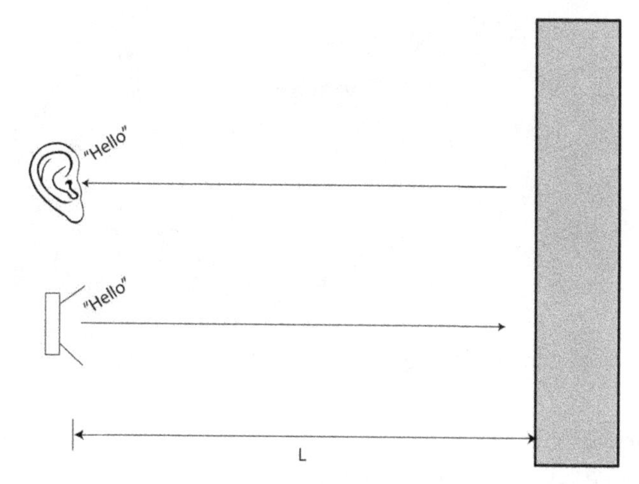

In this figure, the sound travels a distance of 2L from the speaker to the ear.

Example:

How long is the time delay between echoes from a canyon wall 50 m from you at 25°C?

v_s = 344 m/s + 0.6 (25 – 20) m/s = 347 m/s

t = (2 x 50 m) / (347 m/s) = 0.29 s

Easy enough! Just don't forget to multiply the distance to the reflecting wall by two.

Wave Interference and Beats

Wave interference is an extension of the concept of superposition (waves simply add). Interference can occur when two waves cross paths and add together constructively, producing **constructive interference**. Alternatively, the two waves can cancel each other out, producing **destructive interference**. Interference can happen in physical space where one listening location appears louder than another location.

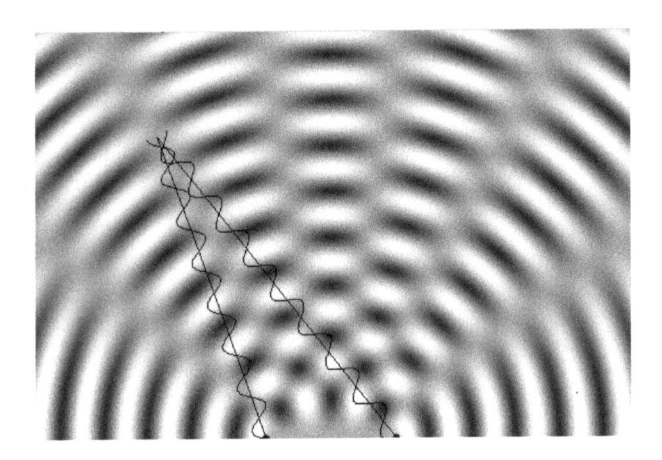

In this figure, circular wave fronts emanate from two sources (at the bottom). In certain locations, constructive interference occurs. In other locations, destructive interference occurs (towards the upper left for example). At any one location, the wave displacement is the sum of the two waves coming from the source. Courtesy of OTTISoft Educational Software.

Spatial interference can be present in certain acoustical situations but typically is not noticeable because musical sounds have many higher harmonics with many different wavelengths. Additionally, at least indoors, there are multiple reflections off the walls that make the situation much more complicated than the simple figure shown above. Actually calculating the locations of either constructive interference or destructive interference can be complicated, even in the simple situation shown in the figure.

Example:

Two sources are generating the same pure tone at 172 Hz. A listener is 20 m from one source and 21 m from the other source. Is this a location of destructive or constructive interference?

$\lambda = (344 \text{ m/s}) / 172 \text{ Hz} = 2 \text{ m}$

difference in distance = 21 m – 20 m = 1 m

The second wave will be 1 m / 2 m = 1/2 way through its cycle when it interferes with the first wave.

The wave goes from peak to trough in 1/2 its cycle. Therefore, the answer is: this is a location of destructive interference. This is about as hard as these problems can get without the use of trigonometry!

Another common form of wave interference is called **beats**. Beats are recognizable when two tones are played slightly out of tune. This is true for complex tones as well as pure tones. Beats are the result of two waves having nearly the same frequency, resulting in times where the two waveforms constructively add, and others where they destructively add. In the figure below, the second wave oscillates with a slightly higher frequency. Through superposition, at any point in time, the resulting wave (the bottom figure) is the sum of the top two waves.

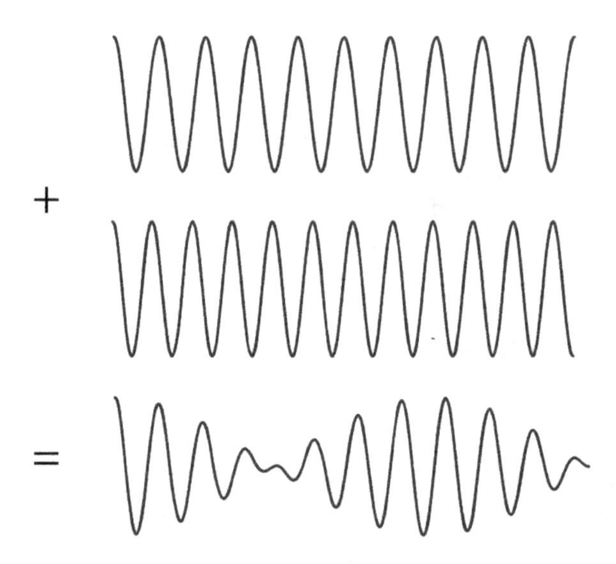

There is a simple formula for the frequency of the beats:

$$f_{beats} = f_2 - f_1$$

where f_2 is greater than f_1.

Example:

Suppose two trumpets are tuning. One plays a note with frequency (frequency of the fundamental) of 440 Hz. The second trumpeter plays a note with frequency of 438 Hz. What is the beat frequency?

f_{beats} = 440 Hz – 438 Hz = 2 Hz.

Beats problems are pretty easy! Just watch out that you don't get a negative number, and if you do, simply flip the sign. When beats get to be more than a few Hertz, we perceive the sound as a flutter or "roughness" as it is sometimes called.

References and Resources

Hall, D., Musical Acoustics, 3rd Edition, 2002

Rossing, T., F. Moore, and P. Wheeler, The Science of Sound, 3rd Edition, 2002

Chapter 4
Harmonic Series

Chapter 4 Outline

- Timbre
- Harmonic Series
- Frequency Spectra
- Spectrograms

Timbre

The pitch "middle C" sounds different when played by different instruments. A flute is easily distinguishable from a piano even when they produce the same pitch. Why is that?

A pitch *may* be a "pure" waveform, consisting of just one single frequency. Most of the time, however, a pitch has a complex waveform that is a blend of many frequencies sounding simultaneously. Each pure frequency within the blend is called a **harmonic**. Each harmonic is special in that it is an "integer multiple" of the **fundamental** frequency. An integer multiple is an integer – like 1, 2, or 3 – times another number. So, if the fundamental frequency is 100 Hz, the first three integer multiples are 100 Hz, 200 Hz, and 300 Hz.

Is 400 Hz an integer multiple of 100 Hz? You bet! In fact, you are often hearing contributions from dozens of harmonics simultaneously. So, a piano string, producing an A with a fundamental frequency of 110 Hz, produces many other frequencies as well: 220 Hz, 330 Hz, 440 Hz, 550 Hz, 660 Hz, and so on.

The pitch we perceive corresponds to the fundamental frequency. The presence of the other frequencies produces the "quality" or **timbre** of the pitch we perceive. The timbre is what distinguishes the sounds of a piano and a flute that are both producing the same pitch, the same fundamental frequency. Each instrument produces a unique mixture of harmonics with different strengths. The various contributions from the various frequencies are a significant part of what produces the *timbre* of the sound – timbre being a word that catches all the perceived qualities of sound that are not pitch or volume. Timbre is also controlled by the manner of sound growth and dissipation, known as transients. Transients include attack, decay, sustain, and release.

Harmonic Series

A guitar string has many ways in which it can vibrate. The different manners of vibration are called **natural modes**. The simplest natural mode on a guitar string corresponds to the largest wave that fits on the string. You could see the **first natural mode** of a guitar string with a strobe light. The strobe light "burns" several snapshots of the string on your retina temporarily. Each different grey line shows the guitar string at a slightly later time. The arrow shows the direction of motion of the string.

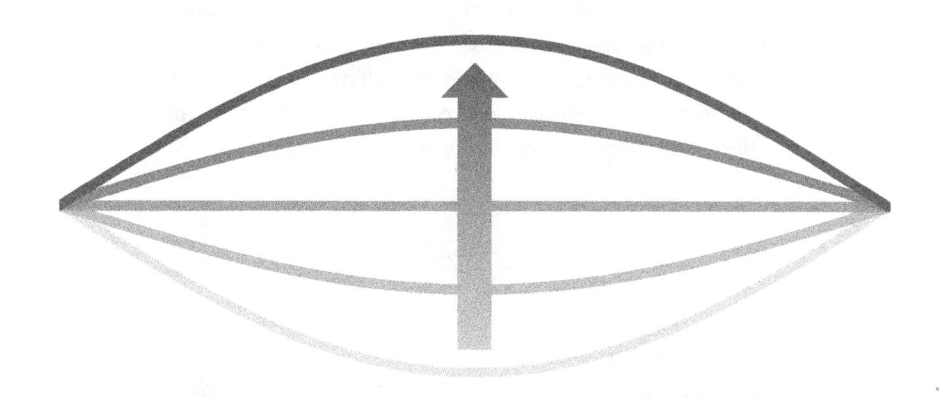

The **second natural mode** looks like this.

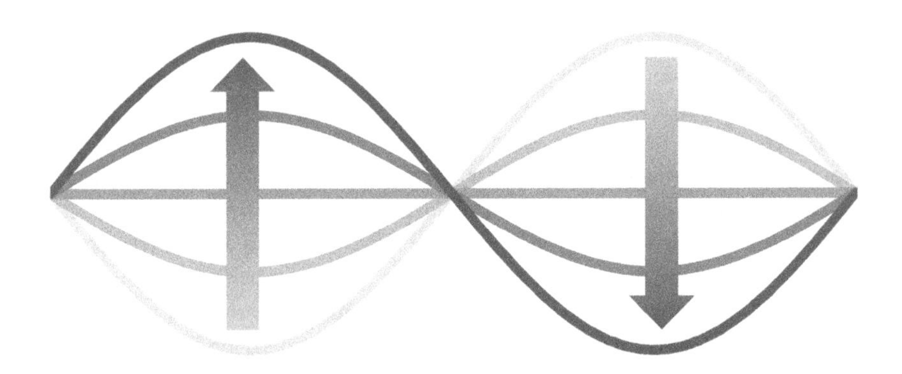

If you cover up the right half of the second natural mode, the left half looks just like the first natural mode except that it's half as wide. The wavelength of the second natural mode is half the wavelength of the first natural mode. These two modes have different wavelengths and therefore different frequencies. When you halve the wavelength, you double the frequency.

$$\lambda \Rightarrow \tfrac{1}{2}\,\lambda$$

$$f \Rightarrow 2\,f$$

A plucked guitar string vibrates with many natural modes. The sound you hear is produced by roughly thirty natural modes, each producing its own frequency. Here are the shapes of the first five natural modes. (To show the shape of each mode, we've drawn only two lines this time, and we've omitted the motion arrows for the fourth and fifth modes for clarity.)

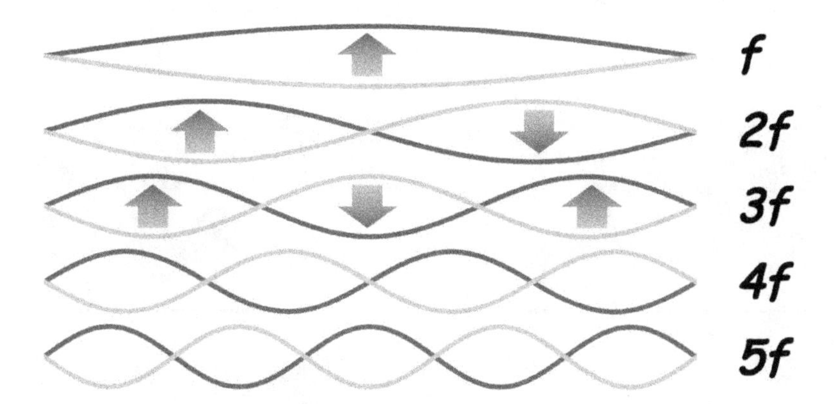

The frequency, f, of the largest wave is called the fundamental or **first harmonic**. The frequencies of the other natural modes form a series with a special pattern. The second mode has a frequency of twice the fundamental, 2f, and this frequency is the **second harmonic**. The third mode has a frequency of three times the fundamental, 3f, and is called the **third harmonic**. The higher modes continue this pattern. All together, these frequencies form a **harmonic series**. The higher harmonics are also called **overtones**. The use of harmonic and overtone can be a little confusing. The first harmonic is the fundamental and is not an overtone. The second harmonic is the first overtone, and the third harmonic is the second overtone.

If an instrument produces frequencies that are part of a harmonic series, the waveform is periodic, and the instrument produces a distinct pitch. Many instruments, however, produce frequencies that are not part of a harmonic series, so the waveform is aperiodic, and there is no distinct pitch. Examples of such instruments include a snare drum, a cymbal, and a gong. We'll talk more about aperiodic waveforms and percussion instruments in Chapter 10.

Frequency Spectra

A **frequency spectrum analyzer** is an important tool to diagnose and understand sound. Most digital sound editing software tools have the capability to do a spectral analysis and produce a **frequency spectrum** – the plural form of spectrum is spectra.

For example, let's say we have recorded a sound wave that looks like this:

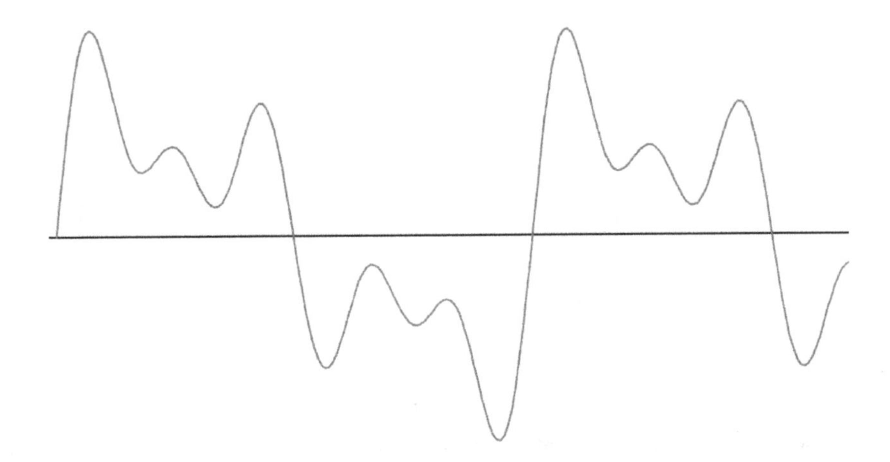

A spectrum analyzer decomposes this complex waveform into its components and displays the "strength" of the individual components.

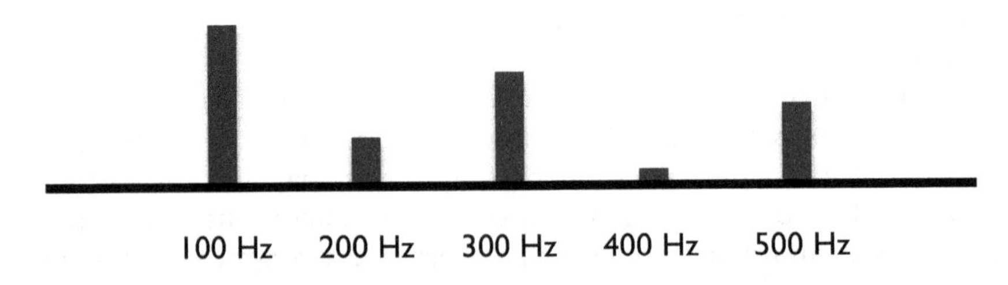

The strength could be displayed as pressure amplitude, intensity, or sound intensity level. These three values are all related to volume, and we'll discuss them in more detail in the next chapter.

Spectrogram

Suppose we want to see how the frequency spectrum changes over time. The frequency spectrum of my voice changes as I sing three descending pitches. We could imagine collecting a whole bunch of frequency spectra, rotating them 90 degrees, and laying them side-by-side in time. This collection of spectra snapshots is a **spectrogram**. Raven Lite is a free software package that produces spectrograms. On the following page, is a spectrogram of one of the authors (JAS) singing in a noisy coffee shop.

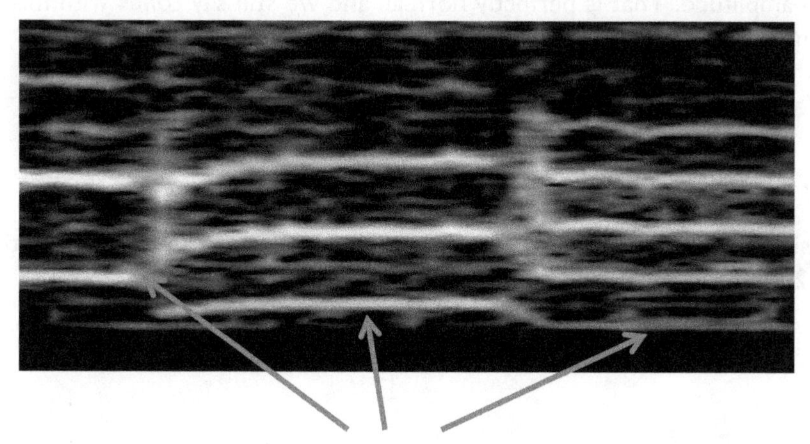

Fundamental

The y-axis is frequency. The x-axis is time. The light intensity in the figure corresponds to the intensity of a particular frequency. The bright white shows high intensity, and black shows no intensity. The white bands are the harmonics produced by my vocal chords as I sing the three pitches. If I were a good singer, these bands would be straighter. You can definitely see me miss my intended pitch on the second note and swoop upward towards the intended pitch ☺ My vocal mistakes are also due to background music interfering with my ability to hear myself. Spectrograms are also called spectrographs or sonograms.

Holding a sung note exactly constant is not necessary for good vocals. In fact, professional vocalists often introduce vibrato that will appear as regular wiggles in the bands as the frequencies fluctuate up and down around the intended pitch.

Pitch and the Harmonic Series

A sound that produces the sensation of a distinct pitch has a distinct harmonic series of frequencies appearing in its frequency spectrum (or spectrogram). The harmonic series is f, 2f, 3f, 4f, … The frequency of the fundamental, f, is the frequency of the perceived pitch. In order for a pitch to be perceived, the tone has to be held steady long enough to complete a cycle or two. This sets a limit on how short notes can be and still be perceived as distinct tones. The relative amplitudes of the various harmonics determine the timbre of the sound. In some cases, various harmonics are missing. For example, in the case of a clarinet, even harmonics have nearly zero amplitude. That is perfectly normal, and we still say tones with missing harmonics are tones that form a harmonic series.

References and Resources

Hall, D., Musical Acoustics, 3rd Edition, 2002

Rossing, T., F. Moore, and P. Wheeler, The Science of Sound, 3rd Edition, 2002

Chapter 5
Sound Intensity and Loudness

Chapter 5 Outline

- Intensity and Sound Intensity Level
- Loudness and its Frequency Dependence

The human ear is an amazing sound sensing device. It can sense an enormous range of intensities. A loud sound that is on the verge of being painful is a trillion (10^{12}) times more intense than the quietest audible sound. Vibrations of the ear drum can be as small as 10^{-8} millimeters (mm) -- smaller than a hydrogen atom!

The frequency range for hearing varies greatly between individuals. For simplicity, we often say that the audible range spans 20 Hz to 20,000 Hz. The ear's sensitivity, however, varies significantly with frequency. In general, we hear low frequencies less effectively than high frequencies. Frequencies below 30 Hz are hard to perceive. The human ear is most sensitive to frequencies between 2,000 and 5,000 Hz, and above 10,000 Hz, our ability to hear declines.

Despite the fact that we say that the upper edge of the audible range is 20 kHz, a healthy young person usually can only hear up to 18 kHz. Our ability to hear high frequencies declines with age, and the decline is much more pronounced in men than in women. By age 55, men typically can't hear frequencies above 5 kHz. Likewise, women lose their ability to hear frequencies above 12 kHz. It is interesting to consider, for example, the impact on the perceived timbre, for older men. The lower harmonics of a complex tone become more important, and they may not perceive subtleties of the amplitudes of the higher harmonics that a younger listener might.

Sound waves carry energy. A sound source radiates energy just like a light bulb or a toaster oven. Energy is measured in **Joules**. **Power** is the rate of energy emission. Power is measured in **Watts** (W), which is Joules per second. If a certain amount of energy is released slowly, the power is lower than if that same amount of energy is released quickly.

Intensity and Sound Intensity Level

Intensity is the amount of power passing through a surface *divided* by the area of that surface. It is measured in Watts per square meter (W/m^2). If one Watt of sunlight is striking a square inch on the ground, the intensity is much lower than if that *same* Watt is focused with a magnifying glass onto a spot 1 mm across. Ants don't burn in the summer sun, but they sure do crackle when under the focused beam of a magnifying glass. (Only a bad person would do such a thing. You are not a bad person, are you?)

Intensity is a measure of the physical strength of sound waves. **Loudness** is our psychological perception of sound strength. Loudness depends both on intensity and frequency. Given two sounds of different frequency and the same intensity, the higher frequency usually will be perceived as being louder than the low frequency.

As a reminder, intensity is measured in W/m^2. The quietest perceivable sound has an intensity of roughly one-trillionth of a Watt per square meter, 10^{-12} W/m^2. We call this intensity the **threshold of hearing** and use it as a reference point for comparison of other sound intensities.

For most acoustic purposes, intensity is converted into **sound intensity level** (SIL). Sound intensity level is measured in decibels (dB). The formula for calculating sound intensity level is:

$$SIL(dB) = 10 \log \frac{I}{I_0}$$

where I is the intensity of the sound and I_0 is the intensity of the threshold of hearing, 10^{-12} W/m^2.

Here is a table of sounds, their intensities, and their sound intensity levels (SIL).

Sound	**Intensity (W/m²)**	**SIL (dB)**
Pain!	1	120
Rock concert	0.1	110
	0.01	100
Subway train	0.001	90
	0.0001	80
City traffic	0.00001	70
Conversation	0.000001	60
	0.0000001	50
Library	0.00000001	40
	0.000000001	30
	0.0000000001	20
Pin drop	0.00000000001	10
Threshold of hearing	0.000000000001	0

Why do we bother converting sound intensity into sound intensity level? After all, the logarithm seems to complicate things a bit? Do we have to use logarithms? We'll show you why logarithms and SIL are useful and worth the added pain. Here's a graph of the intensities of some common sounds.

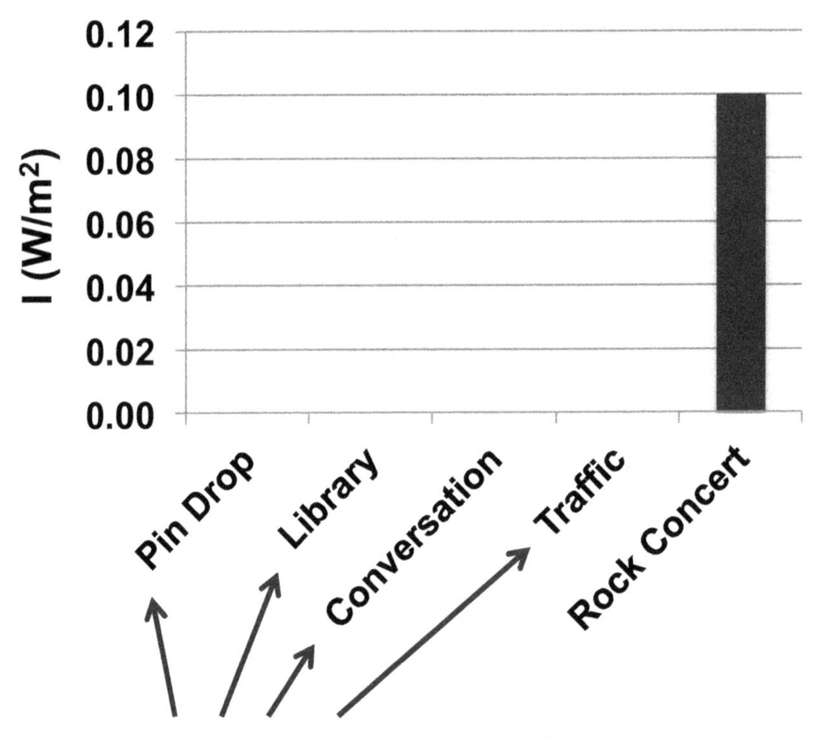

Uh oh! These sounds don't show up very well.

Many things in nature can be counted by ones... one apple, two apples, three apples. There are other things for which counting by one is impractical, like sound intensity.

The sizes of things in the universe is another example. An atom is roughly 0.0000000001 meters across. The nucleus of that atom is roughly 0.000000000000001 meters across. The Sun and the Earth are composed of these tiny particles, but the Sun and Earth are separated by 150,000,000,000 meters. The distance to the next nearest star is 4.4 light years, which is 40,000,000,000,000,000 meters away.

	Size in meters
Diameter of a hydrogen nucleus	0.0000000000000017
Diameter of a hydrogen atom	0.00000000011
Diameter of a living cell	0.00001
Diameter of a human hair	0.0001
Height of a human	2
Diameter of the Earth	13,000,000
Diameter of the Sun	1,400,000,000
Distance between the Sun and Earth	150,000,000,000
Distance between the Sun and Neptune	5,000,000,000,000
Distance to Alpha Centauri	40,000,000,000,000,000
Diameter of the Universe	900,000,000,000,000,000,000,000,000

Logarithms help us to wrap our heads around such enormous disparities in the size of numbers. They allow us to compare the size of the Earth, the Sun, and the Solar System in a manageable way without so many zeros. It is a better way for our head to deal with these very big and very small numbers.

By using a logarithmic scale, we are deciding to count by factors of ten (1, 10, 100, 1000) instead of ones. Likewise, we can count down by factors of ten (1, 0.1, 0.01, 0.001). Here are graphs of the number 5 on a linear scale and a logarithmic scale.

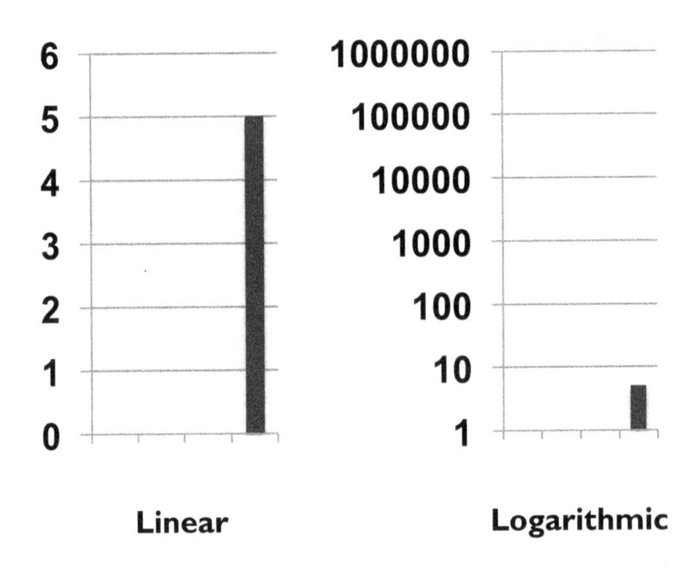

Here are the intensities of various sounds using a logarithmic scale. See! Now, we can picture the intensities of all the sounds on the same graph. When we use a logarithmic scale – in other words, we count by factors of ten – the small values are not *totally* swamped out by the large values. Also, loudness is more closely aligned to the logarithmic scale we are discussing here, and we will discuss this fact further. A sound that is twice as intense is not perceived as twice as loud.

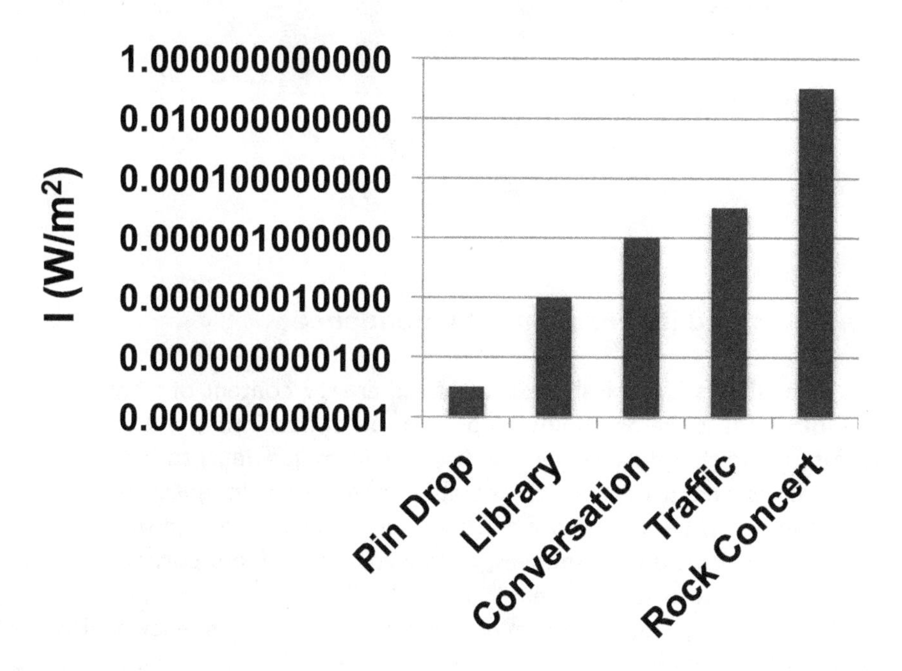

Even though we can see all of the sounds, writing all of those zeros is cumbersome. Rather than graph sound intensities in Watts per square meter, the decibel scale is more convenient. The next graph shows intensity and sound intensity level in units called bels (B) and decibels (dB), named after Alexander Graham Bell even though "bel" is spelled differently.

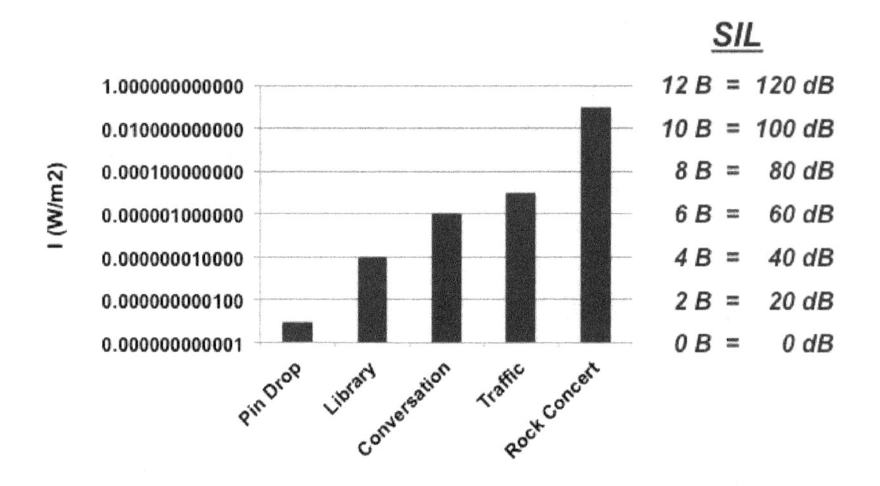

Loudness and its Frequency Dependence

Intensity is a physical measure of the energy content of a sound wave, and **loudness** is the psychological perception of that intensity. Loudness is a sensation and therefore subjective. It varies from individual to individual, but acousticians and psychologists are able to measure loudness by asking a large number of people, "How much louder is this sound than that?" From a collection of individual responses, a doubling of loudness corresponds to a 6-10 dB increase in sound intensity level.

Loudness is partially determined by intensity, but frequency is also very important in producing a particular loudness. This fact should not be surprising since your ear has a limited frequency range. You might suspect your ear would lose sensitivity as the frequency of a sound approaches the lower limit or higher limit of your ears' detection range.

The relationship among loudness, intensity, and frequency is rather complex. This relationship was first measured by Fletcher and Munson in the 1930s. Fletcher and Munson defined a unit of loudness called the phon. One **phon** of loudness is perceived when a pure tone with 1 dB of SIL is produced at 1000 Hz. Likewise, 100 phons of loudness corresponds to 100 dB of SIL at 1000 Hz.

(Sometimes the physical intensity of sound is expressed in terms of what is called a Sound Pressure Level (SPL) that can vary slightly from the SIL depending on the background air density, but we won't worry about this detail, and we will assume the SIL and SPL are very similar – intensity is related to the square of pressure fluctuations. See Rossing (2002) for a detailed discussion of the SPL.)

When other frequencies are produced, the ear's sensitivity is different than it is at 1000 Hz. How loudness varies with both sound intensity level (SIL) and frequency (Hz) is best described using the so-called "equal loudness contours," often called the "Fletcher-Munson Diagram."

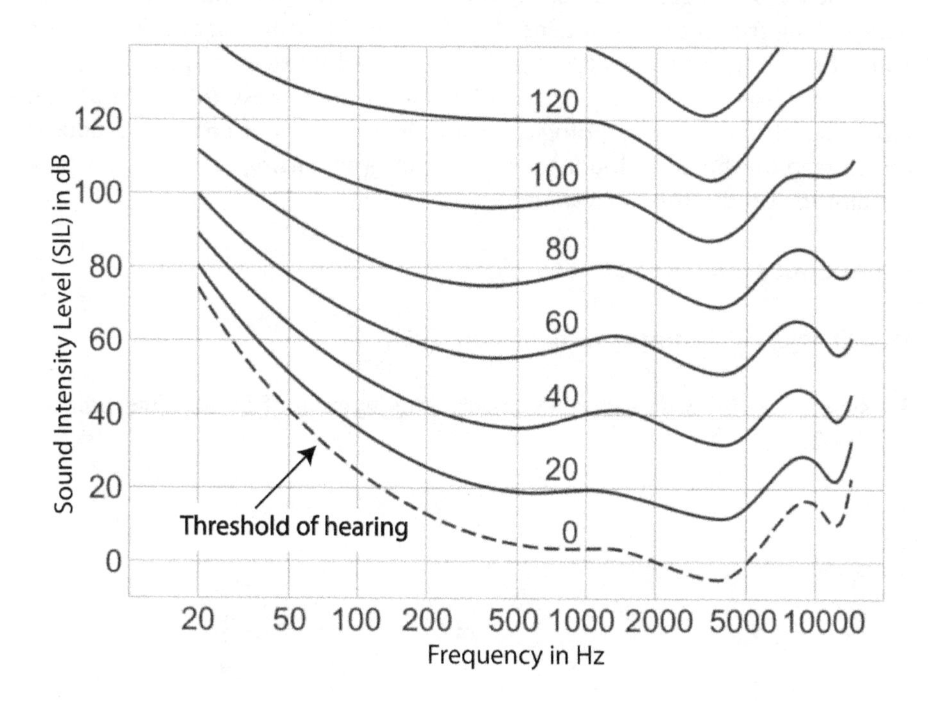

If the ear is highly sensitive to a frequency – say 4000 Hz – then the perceived loudness will be louder ⇒ more phons are perceived than the number of SIL dBs present. A sound with a frequency of 4000 Hz and an SIL of 80 dB produces a perceived loudness of 90 phons. Likewise, at a frequency where the ear is rather insensitive – say 30 Hz – the perceived loudness will be quieter ⇒ fewer phons are perceived than the number of dBs present. A sound with a frequency of 30 Hz and an SIL of 80 dB produces a perceived loudness of 40 phons.

The equal loudness contour diagram is a little tricky to read, but you'll get it with a little practice. Contours of equal loudness are drawn on a "map" of SIL and frequency. The two dips in the contour lines at 4000 Hz and 12,000 Hz show that the ear is particularly sensitive at these frequencies. These frequencies are *resonant* with the natural modes of vibration of the air column present within your ear canal – we discuss the ear canal in Chapter 6 and natural modes of vibration of air columns in Chapter 9.

The loudness versus both SIL and frequency is similar to the altitude (height) versus longitude and latitude on a topological map. Maybe you have seen such a U.S. Geological Survey topological map while planning a hike or backpacking trip in the mountains. The Fletcher-Munson diagram is a similar type of graph, just with a little less familiar x and y-axes. Frequency (x) and SIL (y) are the two axes, instead of position East-West (x) and position North-South (y) on the topological map. The contours of constant loudness are plotted on the equal loudness contour diagram analogous to contours of the altitude on the topological map.

References and Resources

Hall, D., Musical Acoustics, 3rd Edition, 2002

Rossing, T., F. Moore, and P. Wheeler, The Science of Sound, 3rd Edition, 2002

Chapter 6
Hearing Musical Sound

Chapter 6 Outline

- Anatomy of the Human Ear
- The Cochlea: Your Biological Spectrum Analyzer
- Binaural Localization
- Psychoacoustics
- Masking
- Virtual Pitch

Anatomy of the Human Ear

The human ear is an amazing sound detector. It is much more versatile than any microphone. **Dynamic range** is a measure of the spread of intensities that a device can detect. The dynamic range of the human ear is 120 dB, which corresponds to a factor of one trillion between the quietest audible sound to the loudest sound, one that would cause physical pain to the listener. The ears are able to perform this complex function within a very compact space inside the head.

The ear is divided into three regions. The **outer ear** consists of the pinna and the auditory canal. The **pinna** is the curved, folded form of flesh that you see on the side of someone's head – what we usually referring to when we say "ear." The pinna funnels high frequency sound into the auditory canal. The **auditory canal** is the tube through which sound waves travel on their way to other parts of the ear. The auditory canal amplifies frequencies near 4 kHz due a resonance of the ear canal. This resonance makes the ear particularly sensitive to frequencies near 4 kHz.

The **middle ear** consists of the tympanic membrane and the ossicles. The **tympanic membrane**, also known as the eardrum, vibrates because of the pressure fluctuations of the incoming sound waves. The **ossicles** are three small bones called the malleus, incus, and stapes. The more common names for the ossicles are the **hammer**, **anvil**, and **stirrup**. The ossicles act as a system of levers that amplify the magnitude of the vibrations of the tympanic membrane, and they transmit these amplified vibrations to the inner ear. When the sound intensity level is larger than 90 dB, there are two small muscles that pull the ossicles away from the inner ear to help protect it. This protective action is called the **acoustic reflex**.

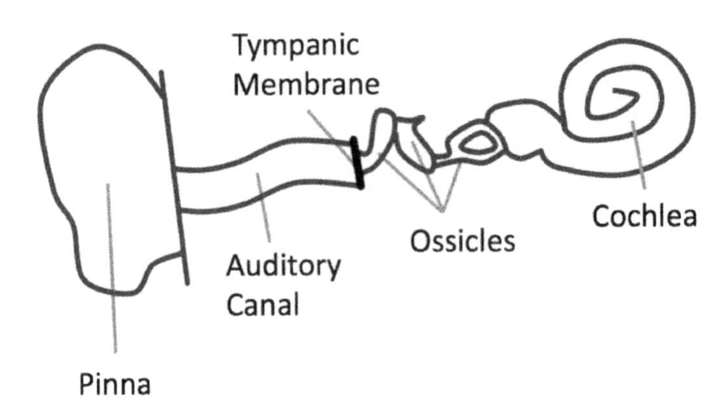

The **inner ear** consists of the cochlea and the auditory nerve. The semicircular canals are also part of the inner ear, but they are not involved in hearing, so we'll ignore them. The cochlea is a small tube that is coiled up into a snail-shaped object the size of a pea. Uncoiled, the cochlea is about 3 cm long. The **cochlea** is an amazing, complex organ that converts the mechanical vibrations of the ossicles into electrical signals that are carried by the auditory nerve to the brain.

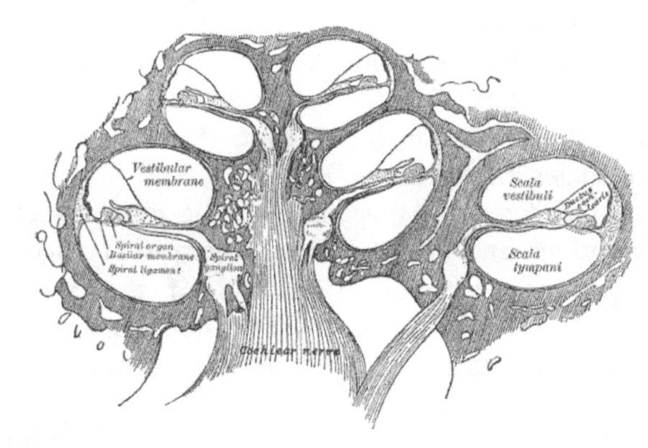

The cochlea is pea sized and coiled up like a snail shell. The hair cells ride on top of the basilar membrane. To the left is a picture from Gray's Anatomy, Henry Gray (1918). This is the view if you sliced the cochlea in half.

The Cochlea: Your Biological Spectrum Analyzer

To illustrate the functioning of the cochlea, let's unwind it.

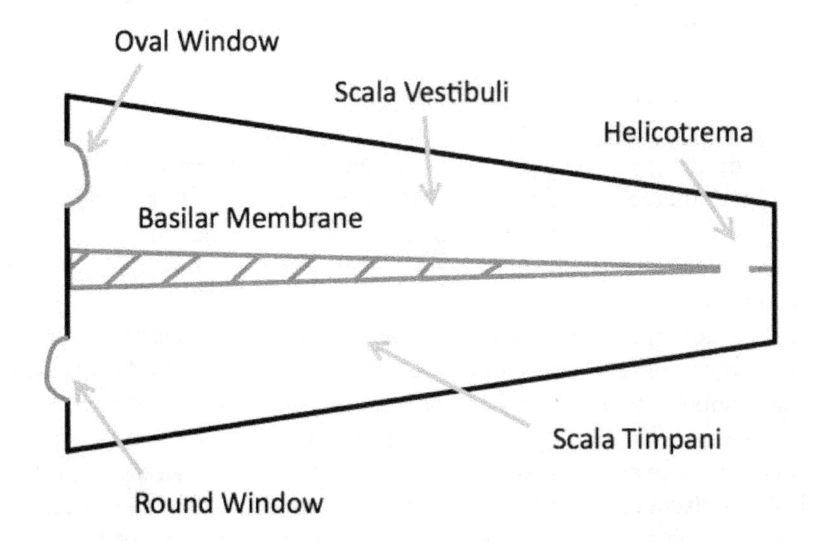

The cochlea is a tube filled with a fluid called perilymph. There are two windows on the ossicle-side of the cochlea. The vibrating stirrup pushes on the **oval window** of the cochlea. The oval window pushes perilymph along the duct called the scala vestibuli, through a hole called the helicotrema, and along another duct called the scala timpani. At the end of the scala timpani, the **round window** bulges outwards, and then the perilymph is reflected back along the path it came. Between the fluid-filled ducts, lies the **basilar membrane**. The basilar membrane has interesting mechanical properties which permit it to act as a frequency analyzer, so that your ear can distinguish the different frequencies it is receiving.

The basilar membrane is more rigid at the end near the oval window and more flexible at the other end. Rigid objects vibrate at higher frequencies than flexible objects. Thus, the rigid end of the basilar membrane resonates when high frequency waves travel the cochlea, and the flexible end of the basilar membrane resonates when low frequency waves travel the cochlea.

Hair cells are nerves with small hairs. There are hair cells all along the basilar membrane, and when a particular region of the basilar membrane vibrates, those hair cells fire electrical signals to the auditory nerve going to the brain. The hair cells residing on different regions along to the basilar membrane are responsible for hearing different frequencies.

When sound waves vibrate the hair cells, the cells send electrical impulses periodically. As the intensity of the sound increases, the hair cells begin "firing" more frequently. So, quiet sounds result in infrequent nerve firing, and loud sounds result in frequent nerve firing.

The vibration of the hair cells cause the hair cells to become active and go in motion, leading to a biological positive feedback mechanism that increases the net frequency of the nerve impulses. The motion of the hair cells can be detected with a special medical microphone. Thus, scientists can test the hearing of a newborn infant by sending a sound into the ear and detecting the motion of the hair cells. These observed hair cell vibrations are called **otoacoustic emissions**.

The various spatial locations of vibrating hair cells along the basilar membrane determine the perceived frequencies of sound. The firing rates of electrical impulses from the hair cells determine the perceived volume of each frequency. Finally, the **auditory cortex** in the brain performs additional signal processing that results in the final overall perception of sound. We'll discuss the functioning of the auditory cortex in a little bit.

You might be thinking that real complex periodic waveforms have many harmonics present and that there are multiple sounds being heard simultaneously. Additionally, the vibrational motion of the basilar membrane and the patterns of electrical impulses traveling along the auditory nerve must be very complicated for a real musical sound. Your thought is completely correct! In listening to a musical composition, many regions of the basilar membrane are excited simultaneously and with different amplitudes. The brain is sent this enormous pile of information and must make sense of it, and the brain only has a fraction of a second to do all this work.

How do we really process the sounds we hear? Do we process the sound in the cochlea or in the brain? The answer is both. As discussed before, the cochlea is a mechanical spectrum analyzer, and spectral information is sent to the brain. The brain's processing is not well understood, but we will see much advancement in this area as magnetic resonance imaging (MRI) is now more commonly used to study the brains of subjects while they are listening to sound.

There is evidence that nerve impulses do fire in unison, carrying transient time information as well as spectral information. This is possible for frequencies lower than about 1000 Hz, where the period of the wave is longer than the shortest time interval between neural impulses. It takes time for nerve cells to recover before being able to fire again. This sets a limit on time resolution of neural signals. There are a few interesting phenomena which support the fact that the brain does quite a bit of signal processing of the information sent by the cochlea.

First of all, there are **binaural beats**. If you play two slightly different tones to each ear, your brain will create the **interference beats** you would hear if the two tones were interfering. The two tones can't really interfere because the two tones aren't present together in the air, a different tone is sent to each ear. Secondly, virtual pitch can work even if higher harmonic tones of the missing fundamental are played in different ears. Virtual pitch is explained at the end of this chapter, but it is a situation where the ear perceives the pitch of the missing fundamental when presented with a few higher harmonics.

On the other hand, the perceptual effects of critical bandwidth and masking clearly show the importance of the cochlea in spectral analysis. **Critical bandwidth** involves the phenomenon where two pure tones that are close in frequency are difficult to distinguish. Critical bandwidth is discussed in this chapter under the heading "Masking" and will also be discussed in Chapter 7. **Masking** is when a louder sound covers up (or masks) a quieter sound that is nearby in frequency. We'll say a little more about masking later in this chapter.

Binaural Localization

Humans and other animals use the sound arriving at two different ears to locate the source. Some animals are extremely adept at localizing sound. A barn owl can localize the sounds of a mouse to acquire its prey in darkness. In fact, the military intelligence community is interested in how a barn owl learns how to do this. Below 1000 Hz, we can localize the source by differences in the timing between the waves arriving at different ears. For frequencies between 1000 and 4000 Hz, our ability to localize sound declines. At even higher frequencies, there is little diffraction (or bending) of the sound waves around the head, and the head casts a sound shadow, resulting in a difference in sound intensity of 10 to 20 dB between the two ears, allowing us to locate a sound source. Also, the shape of the pinna (or outer ear) allows us to identify whether a source of sound is above, below, in front or in back the head for these higher frequencies.

Psychoacoustics

The field of psychoacoustics involves determining the relationship between physical qualities of sound (or stimuli) and the sensations perceived within the brain. How do we relate physical qualities to psychological sensation? Some of the relationships are simple, and some are more complex. Let's look at some of the sensations and associated stimuli in musical acoustics.

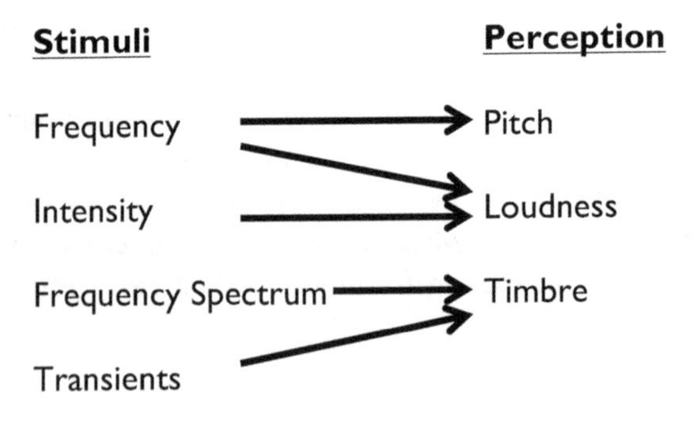

One simple relationship is that between frequency and pitch. It turns out, that pitch depends ever so *slightly* on intensity. A more complicated relationship applies to loudness, where loudness depends on both intensity and frequency, as was shown in the equal loudness contour diagram in Chapter 5. Timbre, being the collection of *all* sound perception that is neither pitch nor loudness, is quite complicated. The frequency spectrum is largely responsible for timbre and your ability to distinguish a clarinet from a trumpet. **Transients** are changes in the waveform with time. How the sound builds (the attack) and how the sound dies away (the release). Transients turn out to be important for timbre. Frequency and intensity may affect timbre as well. Our perception of an instrument's timbre is also affected by the acoustics of the room or location, and even visual queues.

Masking

Masking is when a louder sound masks a quieter sound that is close in frequency. If you look directly at the Sun – (don't do this!) – you are blinded temporarily, and all light colors are masked. The ear is subtler in the way masking takes place. First of all, your hearing recovers much more quickly after hearing a loud sound. Also, lower frequencies tend to mask higher frequencies, but not the other way around. We call the perceived sound the "masker" and the unperceived sound the "source" or "test" tone.

Masking has the following general properties:

1) Lower frequencies mask higher frequencies better
2) The greater the intensity of the masker, the more masking of the source
3) If the two sounds are widely separated in frequency, then little masking occurs

Lower frequencies mask higher frequencies because of the structure of the basilar membrane. Low frequency waves can be resonant with the high frequency regions of the basilar membrane, but high frequencies do not excite regions further down the basilar membrane.

The frequency difference where masking becomes ineffective is when the source is outside of what is called the critical bandwidth surrounding the masker. The critical bandwidth is a measure of how well the cochlea can distinguish two pure tones with different frequencies. The critical bandwidth is approximately 15% of the masker frequency, but depends on frequency.

You might be surprised that the critical bandwidth is so large. After all, you can distinguish two different notes that are mistuned by roughly 0.1%. The critical bandwidth applies to pure tones, which have no higher harmonics present. Musical notes are rich in harmonics, and with the harmonics present, you are much more sensitive to mistuning than you are when two pure tones are sounding.

Masking not only occurs if two sounds are played simultaneously. It can occur for quiet sounds coming either before or after a loud sound. **Forward masking** is a more obvious effect. It occurs when the masker is played a short time before the source – say less than 30 milliseconds. This is related to the physiological recovery time of the hair cells that have recently been excited by the masker. They do not respond as well as rested hair cells. **Backward masking** is when the source precedes a loud masker a few milliseconds later. Backward masking is a little strange. This effect is due to sound processing occurring in the brain. The more intense stimulus interferes with the ongoing processing of the recently received weaker stimulus. Finally, **central masking** is when a source in one ear is masked by a masker in the other ear. This phenomenon occurs in the brain since each cochlea operates independently within each ear.

Background "white noise" can mask sounds, and this is often used in architectural design. White noise generators can be used outside of a closed office door to enhance the privacy of a conversation inside. Those outside the office have more difficulty understanding speech leaking through the door. Another use of white noise involves the relaxing effect of a babbling brook. Such a sound can be used to reduce the distractions from outside road noise.

Human hearing is quite remarkable and can counteract masking. There is something known as the **cocktail party effect**. In a noisy room with many conversations, a person can comprehend a conversation taking place a few feet away. How does the brain do this? The brain uses the fact that the background noise signal is strong in both ears, but the source is stronger in one ear, due to its location. The brain can tune out the noise coming from both ears and focus on the source, which is stronger in one ear. This is a very interesting effect showing how important the brain is in our perception of sound.

Virtual Pitch

Virtual pitch is a subtle and remarkable perceptual effect. One common way to state the concept of virtual pitch is that *"the ear fills in the* *missing fundamental.*" Mathematically, the ear perceives the pitch to be the greatest common factor of the frequencies. Again, this is a conceptually subtle effect. Let's examine a concrete example. If you play a pair of pure tones with frequencies 800 and 1000 Hz, and a second pair of pure tones with frequencies of 750 and 1000 Hz.

Pair 1: 800 Hz and 1000 Hz

Pair 2: 750 Hz and 1000 Hz

How do you think the pitch changes? Does the pitch go up or down? Well, you might guess that the perceived pitch goes down since 750 Hz is less than 800 Hz, but this turns out to be completely wrong. The reason is the psychoacoustic effect of virtual pitch.

The frequency of the lower pure tone within each pair does goes down from 800 Hz to 750 Hz. However, we perceive the pitch going up. Why? Because of the effect of virtual pitch. The greatest common factor is 200 Hz for the first pair of pure tones and 250 Hz for the second pair. The brain perceives pitch as the greatest common factor of the frequencies present – the brain "fills" in the missing harmonics that it is *expecting* to accompany the frequencies that are actually present. This is virtual pitch. So, the perceived pitch goes up from 200 Hz to 250 Hz. Neither 200 Hz nor 250 Hz are present in either of the pairs of pure tones. It is quite remarkable the brain does this. It seems rather complex and is certainly hard to do in our heads quickly. Let's look at another example:

Pair 1: 400 Hz and 800 Hz (we'd perceive a 400 Hz pitch)

Pair 2: 800 Hz and 1200 Hz (we'd perceive a 400 Hz pitch!)

So, in this example the perceived pitch of the two sounds is the same. This is a surprising result. If you find virtual pitch to be something strange, you are not alone. Virtual pitch is an active area of research in the speech and hearing sciences. Virtual pitch is important in a variety of acoustical contexts. It is important in communications where lower frequencies do not necessarily need to be transmitted or received and can be filtered out to reduce the amount of information (called the bandwidth). For example, it is used in the old-fashion touch-tone phone, where the perceived pitches/frequencies of the touch-tones are not actually transmitted. It is also important in instruments. Often the perceived pitch is at a frequency where the instrument has little or no harmonic intensity. For example, the lower notes on the acoustic piano.

References and Resources

Hall, D., Musical Acoustics, 3rd Edition, 2002

Pierce, J., The Science of Musical Sound, 1992

Rossing, T., F. Moore, and P. Wheeler, The Science of Sound, 3rd Edition, 2002

Chapter 7
Musical Intervals and Scales

Chapter 7 Outline

- Numbers and Notes
- Relationship among Notes and the Piano Keyboard
- Beyond the Chromatic Scale
- The Major Scale
- The Pythagorean Hypothesis
- Pythagorean Tuning
- The Pythagorean Comma
- Just Tuning
- Cents
- Critical Bandwidth
- Consonance and Dissonance

Numbers and Notes

In this chapter we further explore the frequency relationships among musical notes. We'd like to understand why modern day notes are assigned specific standard frequencies and why there are the twelve **semitones** (or **half steps**) within an **octave**. In addition, we will develop a better understanding of why certain **intervals** are consonant and others are dissonant. As you may already know from reading Chapter 6, pitch and frequency are distinct entities. **Pitch** is a psychological perception, and **frequency** is a physical quantity we can measure with instrumentation. The sensation of pitch depends almost entirely on the **frequency of the fundamental**. Therefore, in this chapter we will often refer to the frequency of a note even though we are really talking about the perceived pitch.

Certain relationships between two or more notes (or pitches) seem natural and just sound "right." These notes, when played sequentially or together are **consonant**. Consonance (and it's opposite - dissonance) combined with rhythmic sounds are a significant portion of what music is. The first understanding of the relationships between frequencies of notes was discovered by the Greek philosopher Pythagoras (580 BC). Pythagoras, Buddha, and Confucius were actually all contemporaries, amazingly enough! They were quite the "new age" thinkers back then. Pythagoras had a cult-like following. Pythagoras believed in "The Music of the Spheres," that the soul must be in resonance with nature, and "all is number." Pythagoras is even credited for inventing the word mathematics (in Greek of course). Unfortunately, there are no written works by Pythagoras. The mysticism and secrecy of Pythagoras and his followers were questioned by the public, and his institute was burned (and apparently so was Pythagoras).

Pythagoras' important contribution to music was figuring out the ratio relationships between musical intervals. It is actually quite a remarkable feat. Pythagoras did not fully understand frequency and wavelength relationships, so his understanding of integer ratios is based on the pitches produced by strings of different lengths. He used an instrument called a **monochord** (scientists prefer the term **sonometer**). The **monochord** is a very simple one-string instrument used for studying of sound.

Pythagoras' important contribution is what we call the **Pythagorean Hypothesis:** that there are simple integer relationships between consonant intervals. We will come back to the Pythagorean Hypothesis shortly. First, we need to set the "lay of the land" by understanding the notes we routinely use today.

Relationships among Notes and the Piano Keyboard

The piano keyboard is laid out to be easy and efficient to play music with our fingers. The piano is a relatively modern instrument, and its keyboard was designed around the structure of music theory. The piano is not constructed with keyholes, valves, and other mechanical devices that result in the unintuitive fingerings involved in playing brass and woodwind instruments.

If you can read music and find notes on a keyboard, you know quite a bit already. For the rest of you, we need to be able to find notes on the keyboard (it is OK if you fumble around at it) to continue with our investigation of musical intervals and scales. The figure below shows a keyboard to familiarize yourself with the notes. The white keys produce the "natural" notes. The black keys produce the sharped and flatted notes. The black key between C and D has two names. Depending on the musical context this black key is called "C sharp" ($C^{\#}$) or "D flat" (D^{b}). The note $C^{\#}$ and D^{b} are the same. Likewise, the black key between D and E which is $D^{\#}$ or E^{b} is the same note. (If you are playing in some of the more exotic musical keys, you may encounter an $E^{\#}$, which is played with the F key. Likewise, a C^{bb} is played with the B^{b} key.)

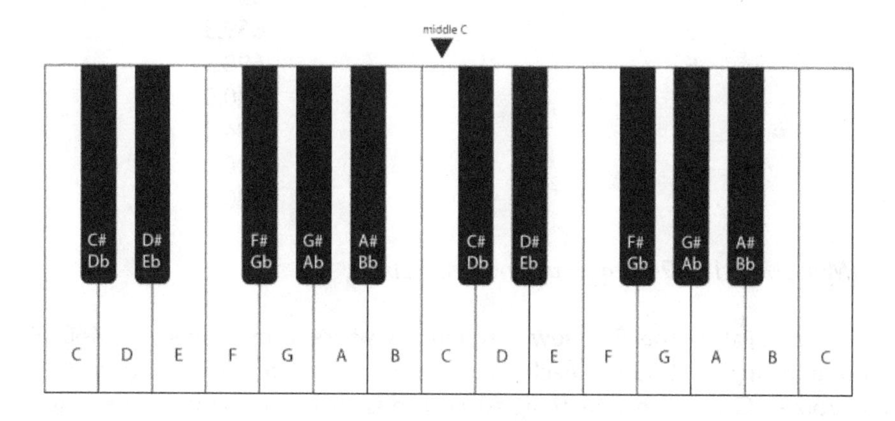

Graphic by Jon Hart

Present day Western instruments use a tuning called **equal temperament**. Concert A is conventionally the reference frequency and defined as 440 Hz. Some orchestras will use a different frequency for concert A, such as 435 Hz or 442 Hz. It's not that important what the actual value of concert A is, just as long as everyone in the ensemble uses the same frequency as their reference. Concert A is the note above middle C on the piano keyboard. As the notes go upwards one semitone (or half step) at a time, the frequency increases. You can calculate these frequencies by multiplying by the 12th root of two ($2^{(1/12)} \approx 1.05946$). The 12 notes in the chromatic scale are given in the following table. The chromatic scale with equal temperament tuning became widely used in the 1700's. For example, J.S. Bach used equal temperament in "The Well-Tempered Clavichord" (1722). The clavichord was a forerunner of the piano-forte that we now call the piano.

Note	Frequency (Hz)
A	440.0
$A^{\#}$	466.2
B	493.9
C	523.3
$C^{\#}$	554.4
D	587.3
$D^{\#}$	622.3
E	659.3
F	698.5
$F^{\#}$	740.0
G	784.0
$G^{\#}$	830.6
A	880.0

(Multiply by 1.05946 to go up one half step.)

Try calculating the first few frequencies yourself by starting at 440 Hz and multiplying by 1.05946 each half step. Check your answers against the table above. One important thing to realize is that our ear is very sensitive when it comes to perceiving pitch. Therefore, we need to be fairly accurate in our calculations and keep at least 4 significant digits for frequency calculations.

Digging a little more deeply, we'd really like to understand the origin of consonant and dissonant musical intervals. We begin with the most consonant of all intervals, the octave. The **octave** has a frequency twice that of the "root" note. The octave is the most fundamental interval. Perceptually, it is the most natural unit of pitch. The octave sounds really, really similar to the root, just higher in pitch. This is probably the most accurate one-to-one mapping between physics and perception. For example, using digital sound editing software you can generate a pure tone at 55 Hz and go up by factors of two and listen to the sensation. 55 Hz, 110 Hz, 220 Hz, 440 Hz, etc. You will hear octave intervals. All cultures use the octave heavily. Human beings appear to be hard-wired for the octave interval.

Beyond the Chromatic Scale

There are musical scales that have more than 12 notes. Carnatic music from the southern parts of India uses a 22-note musical scale. Scales with further subdivision of the octave and more than 12 notes are called "microtonal scales." The music of the late and quite iconic *avante garde* musician Harry Partch (1901-1974) explored microtonal scales extensively. It is worth checking his music out.

The Major Scale

The major scale (also called the diatonic scale for you theory buffs) contains eight notes spanning one octave. There is a specific pattern of seven intervals between the notes of any scale:

H = half step = 1.05946

W = whole step = two half steps = 1.05946 x 1.05946

The major scale follows a pattern determining the frequency interval between notes:

W W H W W W H

Example: The D Major Scale

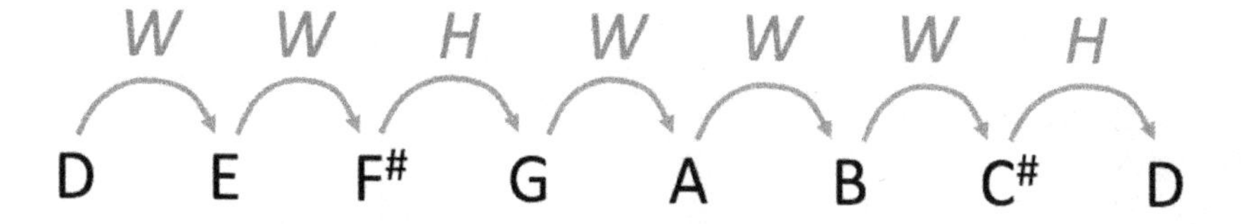

To work through this example, you start on D on the piano keyboard and proceed upwards with the WWHWWWH pattern. If we changed the concert A from 440 Hz to 442 Hz, you could calculate the new frequencies associated with this retuned D major scale. Try it. You should be able to calculate the frequencies of each note regardless of the reference frequency for concert A.

The Pythagorean Hypothesis

Why are there consonant and dissonant intervals? Is there a simple answer to this question? Well, there are quite remarkable mathematical patterns among the most consonant intervals. Pythagoras discovered these patterns over 2500 years ago using a monochord. There turn out to be simple integer relationships between consonant intervals. The Pythagorean Hypothesis is that consonant intervals in music are produced by certain special ratios of frequencies. These special ratios are ratios involving two small integers like 2/1, 3/2, and 5/4. Some important musical intervals you may be familiar with are given in the table below.

Interval	Note	String Length	Frequency Ratio
Root	C	L	1
Minor third	E♭	5/6 L	6/5
Major third	E	4/5 L	5/4
Perfect fourth	F	3/4 L	4/3
Perfect fifth	G	2/3 L	3/2
Minor sixth	A♭	5/8 L	8/5
Major sixth	A	3/5 L	5/3
Octave	C	1/2 L	2/1

Let's examine more closely one of the most consonant intervals. The perfect fifth has a frequency that makes a 3 to 2 ratio with the root.

Examples of fifths

C (root) and G (fifth)
D and A
B and F♯

Pythagorean Tuning

Pythagorean tuning was commonly used in Western culture, up until the 18th century. Both Pythagorean Tuning and just tuning – to be discussed later – are still often used for the pedal harp to create a musical effect. The pianist Michael Harrison uses some of these ancient tunings of the Greeks in his music by modifying a grand piano. For example, listen to the recording by Michael Harrison, "From Ancient Worlds" (1993). In Harrison's work, the piano, at times, sounds like an Aeolian harp.

Pythagorean Tuning is based on perfect fifths. Proceeding up the musical scale by fifths results in creating every note within the chromatic scale.

C ⇒ G ⇒ D ⇒ A ⇒ E ⇒ B ⇒ F# ⇒ C# ⇒ G# ⇒ D# ⇒ A# ⇒ F ⇒ C

Usually these notes are written in a circular pattern, and this circular pattern is called the "**circle of fifths.**"

Note	Frequency Ratio (relative to C)
C	1
G	3/2
D	3/2 × 3/2
A	3/2 × 3/2 × 3/2
etc.	etc.

After proceeding through the entire circle of fifths, we have climbed seven octaves. We can take each note above the first octave and shift it down octave by octave, until it lands within the first octave. In doing this, you create the chromatic scale. Pretty neat, huh?

The Pythagorean Comma

So, there's a problem with this Pythagorean method of creating the chromatic scale by multiplying by 3/2 and dividing by 2. We *try* to create the octave of the root C by multiplying the frequency of F by 3/2, and we *fail!*

$$(3/2)^{12} / 2^6 = 2.027$$

Ugh! We need the octave to have a frequency that is twice that of the root. 2.027 is far enough away from 2 that this octave sounds horrible. This imperfection in Pythagorean tuning is known as the **Pythagorean comma**. The octave is not the only mistuned note. Notes using Pythagorean tuning are spaced a little funny. A song played in different keys will sound different with each key. Some keys will sound OK, and others will sound downright awful.

There are many compromises we can make to adjust the imperfect scale generated using Pythagorean tuning. Keep in mind that if we play music in one key there is really no problem. One improvement over Pythagorean tuning is called just tuning. Equal temperament is the modern compromise we use to fix the Pythagorean Comma. Equal temperament nudges each of the Pythagorean notes up or down a little bit until all the frequency ratios are the same. Half steps are different in frequency by a ratio of the twelfth root of two (1.05946). Whole steps are different in frequency by a ratio of the square of the twelfth root of two (1.12246).

Just Tuning

Building on the Pythagorean Hypothesis (that music is based on integer ratios), we can build the following scale:

C	D	E	F	G	A	B	C
1	?	5/4	4/3	3/2	5/3	?	2

For D, take two perfect fifth intervals up, then drop an octave.

3/2 x 3/2 x 1/2 = 9/8

For B, go up a perfect fifth, then up a major third.

3/2 x 5/4 = 15/8

Now, we have built a major scale! We call this type of frequency assignment just tuning. We can fill in the sharps and flats, or build a chromatic scale, by taking intervals from these notes that give flats and sharps. The choice of interval used in defining a given sharp or flat note will affect the frequency obtained. For example, using the minor sixth interval (8 semi-tones up) from C to obtain A^b, gives a frequency ratio of 8/5.

8/5 = 1.6

But, then requiring E to $G^\#$ or that A^b be a major third interval (which it should be, 4 semi-tones up) gives

5/4 x 5/4 = 25/16 = 1.5625

As you can see, there is an inconsistency between these two intervals. You can see that G# is different than A♭. In fact, organs sometimes have different keys for sharps and flats! We don't want such a complicated situation. This is exactly the problem that equal temperament fixes, although at a musically perceived cost. A slight compromise from integer ratio tuning is required. This compromise is fairly perceivable for the major third interval. Pythagoras would view the equal tempered tuning of the major third interval as being mistuned by 0.8%.

One of the worst mistunings in the just tuning scale is called the diesis, which is up three major thirds and down an octave.

$(5/4)^3 / 2 = 125/128 = 0.9766$ ...an error of 2.3%

For you serious musicians, we are sure you have played a simple melody or pattern in all twelve keys. If your instrument used just tuning, you'd be in for some surprises!

Cents

A **cent** is a very small interval used to measure small differences in tuning. The term often comes up in practical applications (tuning instruments), so it is important to know what it is and how to use it. The unit of cents is often used on tuning meters and also used in discussion of temperament and tuning. It is easy to understand, and when it comes up, if you forget what it is, just look it up in the book!

1200 cents = 1 octave
100 cents = 1 semitone

Like all other intervals, the cent is a multiplying constant. That is, we multiply 100 cents together to get a semitone.

1 cent = $2^{1/1200} = 1.000578$
100 cents = $2^{100/1200} = 2^{1/12} = 1.05946$ = 1 semi tone

That's it, not too complicated.

Critical Bandwidth

If we listen to two pure tones, keeping one frequency fixed while varying the other, we can study the ear's ability to distinguish between the two tones. We can do this using digital sound software to generate two sine waves. You can do this experiment on your own. If the frequencies of the two tones are very close, you will hear beats, which we discussed earlier. The frequency of the beat is the difference between the two frequencies, and the perceived pitch is at the average of the two frequencies.

If you tune one of the frequencies further away from the other you will hear a rough sound, but you will not hear two distinct pitches. Move the frequencies further apart and you begin to perceive two distinct pitches. This region of beats and roughness is what is called the **critical bandwidth**. It is actually a physiological phenomenon, because the cochlea sends different frequency ranges down different nerves. The ability of the cochlea to discern between two frequencies is not arbitrarily precise, and the critical bandwidth is a measure of this limit in distinguishing two similar frequencies. The critical bandwidth is determined using pure sinusoidal tones, and things are quite different with complex musical tones. The pitches of two complex tones are much easier to distinguish because of large differences among the higher harmonics.

The figure below shows how the critical bandwidth varies with frequency. For frequencies above 500 Hz, the critical bandwidth is a little less than a minor third interval. We can think of the critical bandwidth as approximately two to three half-steps.

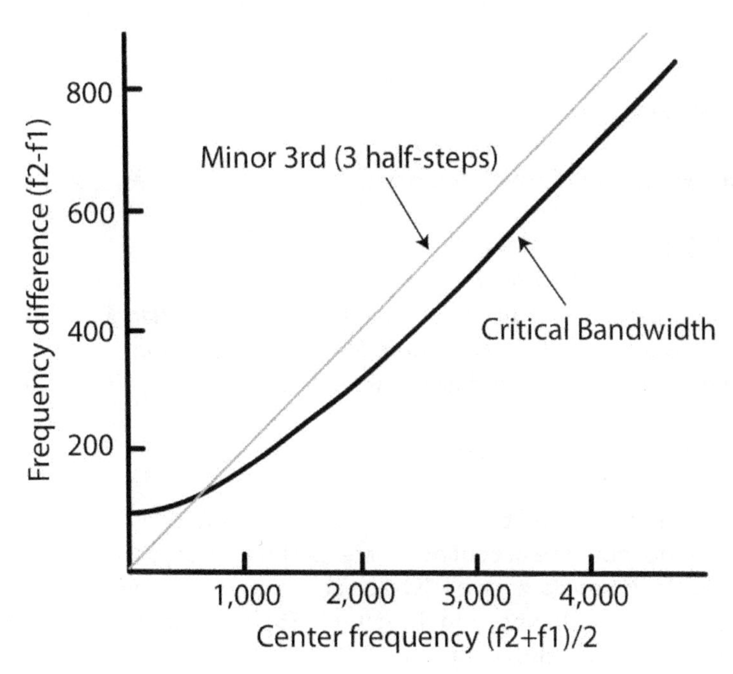

The critical bandwidth (y-axis) is a difference between two frequencies being played at the same time. It is measure of the ear's ability to distinguish between the two sounds. The critical bandwidth depends on the "center frequency" the average of the two frequencies being compared (x-axis).

Consonance and Dissonance

A discussion of critical bandwidth naturally begs the question, "Can we explain something about consonance and dissonance?" Specifically, can we explain why an interval sounds happy, sweet, simple, sad, bitter, or tense? When we use the word dissonance here, we are not implying the sound is unpleasant. In fact, music using only consonant intervals becomes quite dull. It is generally true that more training or listening is needed to appreciate dissonant intervals.

One theory explaining consonance and dissonance, which seems to make the most sense, is that consonance occurs when the harmonics of the two complex tones most closely overlap. It is also important that the lower harmonics that do not overlap, are well separated. This condition avoids beats or roughness if the two complex tones are played together. By well separated, we mean the separation of the fundamental frequencies is larger than the critical bandwidth. Take for example a 100 Hz complex tone having harmonics:

100, 200, 300, 400, 500, 600, ...

An octave up would have harmonics:

200, 400, 600, 800, ...

These harmonics of the two complex tones, overlap perfectly. The octave is an extremely consonant interval. A perfect fifth interval above the root would start at 150 Hz and have harmonics:

150, 300, 450, 600, 750, 900, ...

So, the harmonics of the root and perfect fifth overlap pretty well. Where they do not, the frequencies are fairly well separated. We could compare other intervals as well. Other intervals will show less and less overlap. Researchers Plomp and Levelt (1965) had made an attempt to quantify this, and this is shown below.

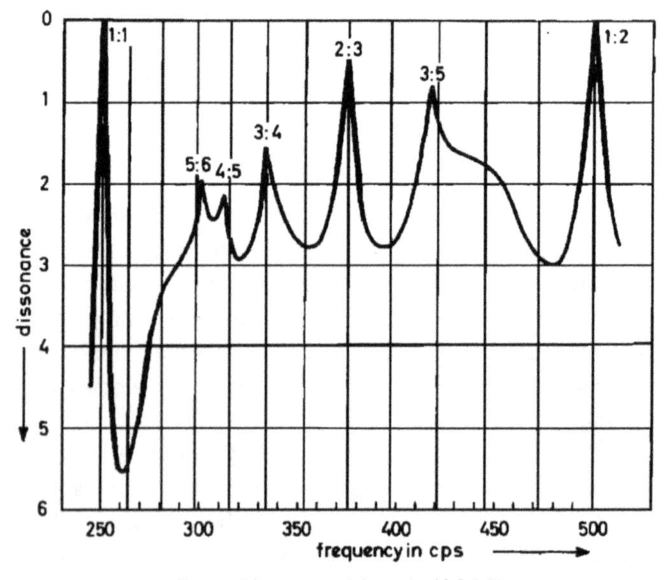

From Plomp and Levelt (1965)

References and Resources

Garland, T. and C. Kahn, Math and Music: Harmonious Connections, 1995

Pierce, J., The Science of Musical Sound, 1992

Plomp, R. and W. Levelt, "Tonal Consonance and Critical Bandwidth," *J. Acoust. Soc. Am.* **38,** 548 (1965)

Rossing, T., F. Moore, and P. Wheeler, The Science of Sound, 3rd Edition, 2002

Zwicker, E., E. Flottorp, and S. Stevens, "Critical Band Width in Loudness Summation," *J. Acoust. Soc. Am.* **29,** 548 (1957)

Chapter 8
String Instruments

Chapter 8 Outline

- Waves on a String
- Standing Waves
- Nodes and Anti-nodes
- Plucked and Hammered Strings
- The Bowed String
- Instrument Body
- Violin Body

Waves on a String

To understand the science of string instruments, we need to start out simple. We'll begin by discussing an ideal string with no stiffness. What do we mean by that? We mean a floppy string that it is perfectly flexible and cannot support itself, like the thin white string in a ball that you might find at the grocery store, also known as butcher's twine. Other strings, like guitar strings, may have *some* stiffness, yet still behave approximately as an ideal string (with no stiffness). Some piano strings have significant stiffness, and this stiffness results in a unique timbre.

One of the simplest stringed instruments is a called a **monochord** or a **sonometer**. A monochord is made by attaching each end of a flexible string to a rigid body.

The monochord can be played by plucking the string. We displace the string transversely – in other words, we pull on the string – and release it. The tension in the string pulls the string back towards its equilibrium position. The momentum of the string causes the string to continue traveling past the **equilibrium position** until the tension becomes great enough to stop the string and reverse its motion, thus creating a vibration.

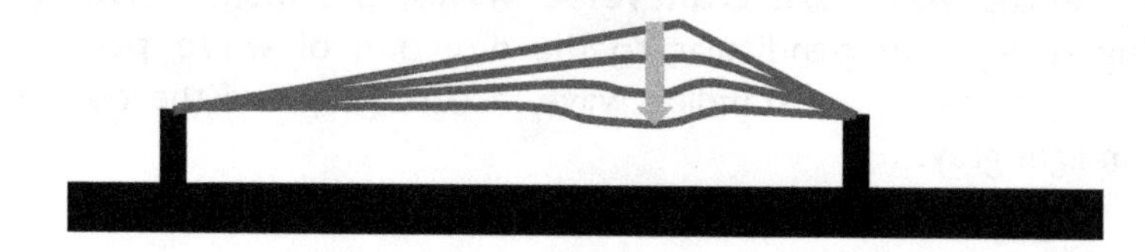

The string's vibrational motion is complex. The complex motion is made up of numerous **natural modes**. The first five natural modes look like picture below.

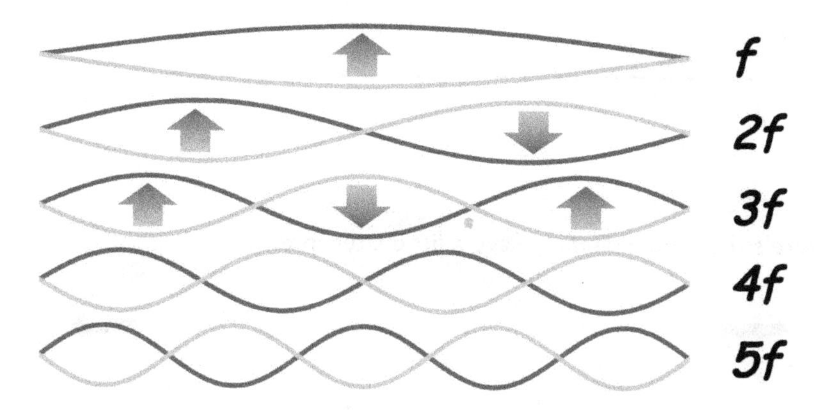

Standing Waves

Each natural mode is a **standing wave**. The wave appears to move solely up and down, but a standing wave is actually made of two waves that are traveling in opposite directions and reflecting back off of the fixed end-points. These two waves traveling in opposite direction have the same amplitude, and from our discussion in Chapter 3, simply add together. These standing waves are **transverse** waves; the **displacement** of the vibrating string is perpendicular to the direction of **wave propagation**. Below, we've drawn the standing wave in dark gray and the two traveling waves in light gray.

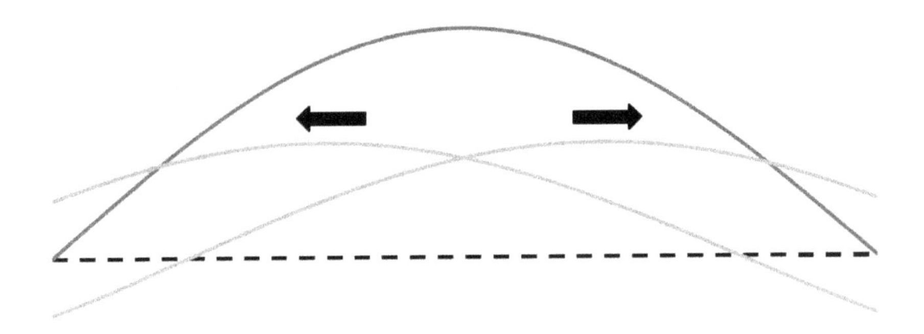

Here's the same standing wave a little later in time.

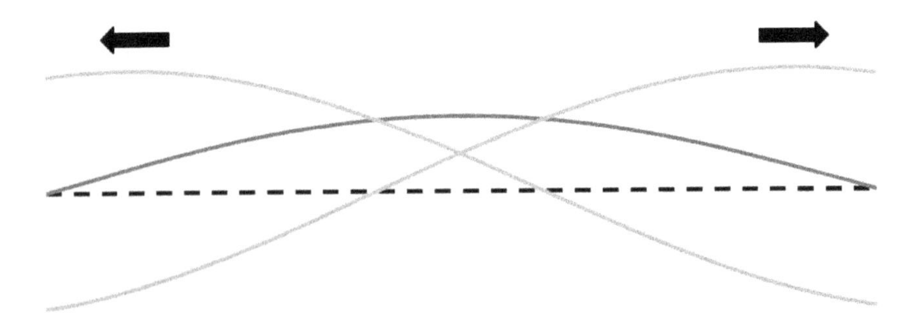

When the waves reflect off of the fixed end-points, they flip upside down – technically, the waves undergo a phase shift of 180° upon reflection from a fixed end-point.

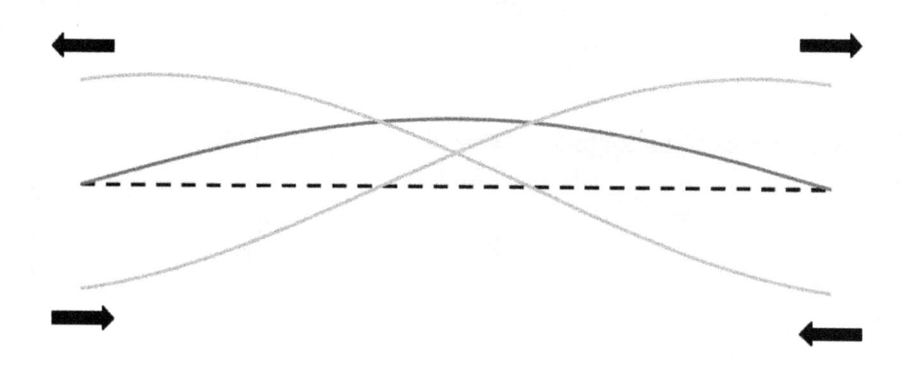

The speed (v) of a transverse wave along an ideal string is:

$$v = \sqrt{T / (m/l)}$$

T is the tension on the string. The tension is a force and is measured in Newtons. "m / l" is the ratio of the mass of the string to the length of the string. You measure the mass and measure the length and divide the mass by the length – it is really that simple. This formula for the speed of a wave on an ideal string with zero stiffness works pretty well for real strings on most stringed instruments.

The **wavelengths** and **frequencies** for standing waves on an ideal string follow a pretty simple pattern.

Wavelength of Natural Mode	Frequency of Natural Mode ($f = v / \lambda$)	
$\lambda_1 = 2L / 1$	$f_1 = v / \lambda_1$	$= 1 f_1$
$\lambda_2 = 2L / 2$	$f_2 = v / \lambda_2$	$= 2 f_1$
$\lambda_3 = 2L / 3$	$f_3 = v / \lambda_3$	$= 3 f_1$
$\lambda_4 = 2L / 4$	$f_4 = v / \lambda_4$	$= 4 f_1$
$\lambda_5 = 2L / 5$	$f_5 = v / \lambda_5$	$= 5 f_1$
...		
$\lambda_n = 2L / n$	$f_n = v / \lambda_n$	$= n f_1$

The standing waves on an ideal string *do* produce a **harmonic series.**

The zero stiffness approximation fails for low-pitched piano strings because they are quite thick and stiff. The stiffness causes the speed of the wave to increase for the natural modes having short wavelengths. This increase in the wave speed increases the frequency of these natural modes. Therefore, these modes are inharmonic, meaning that their frequency is not an integer multiple – f, 2f, 3f, etc. – of the fundamental frequency. Inharmonic modes are called **partials**. We will discuss inharmonic vibrations and partials in Chapter 10 when we describe how percussion instruments work.

The fact that low-pitched piano strings produce inharmonic frequencies is important for tuning a piano. Piano tuning uses a technique called **stretched tuning**. Low notes are detuned slightly so that the fundamental frequency is slightly below the ideal frequency for equal temperament. This is done because the ear is rather insensitive to the low frequencies of the fundamental and the second harmonic. The ear is much more sensitive to the higher frequency partials and perceives the virtual pitch associated with these higher partials. Virtual pitch is the psychoacoustic property where the ear fills in the fundamental when presented a few higher harmonics. The ear perceives the pitch of this "missing" fundamental. Virtual pitch was discussed at the end of Chapter 6.

Nodes and Antinodes

If you look back at the natural modes on a string, you'll see that there are locations on the string that are stationary. These stationary points are called **nodes**. In between the nodes are locations where the string displacement is the largest. These maxima are called **antinodes**. Nodes and anti-nodes are important musically. Varying the position of electric pickups along a string changes the timbre from full and mellow to bright and twangy.

If the pickup is close to an anti-node, that particular natural mode will be captured strongly. If a pickup is close to a node for a particular natural mode, that natural mode will be captured weakly by the pickup. If the pickup is in between a node and an anti-node, that natural mode will be captured somewhat.

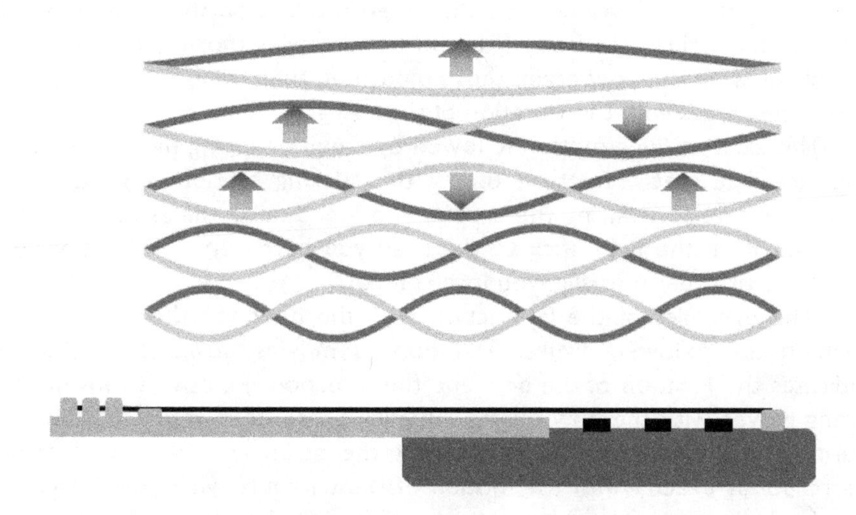

Electric guitars have multiple pickups placed at different locations along the strings. The neck pickup captures the low harmonics more effectively than the high harmonics, producing a warm, mellow timbre. The bridge pickup captures the high harmonics more, thus producing a bright, twangy timbre.

Plucked and Hammered Strings

In reality a pluck can be approximated as an initial triangular wave shape with a rich spectrum of higher harmonics. A hammer striking the string produces a very similar initial wave shape, such as what happens when a acoustic piano key is pressed. The string vibrates the bridge that efficiently couples the vibrational energy to the instrument body. It is the natural modes of the body that we actually hear. Higher harmonics die away quickly because they couple to small scale wiggles in the body which involve a lot of stress and strain in the wood and are dissipated ultimately into heat – the amount of heat is too small for you to feel. Lower harmonics resonate large scale vibrational modes in the body that primarily dissipate through radiating sound waves.

The Bowed String

Bowing a string involves a **stick-slip mechanism**. For most of the cycle, the string sticks to the horsehairs on the bow and is being pulled away from the equilibrium position. When the pull on the string is strong enough, the string slips free of the bow and travels more quickly until the string sticks to the bow again, thus completing an oscillation. This stick-slip action resonates with the vibration of the string.

This slow sticking motion followed by a quick slipping motion is due to the fact that **static friction** during the sticking motion is greater than **dynamic friction** during the slipping motion. Similar things happen when driving. While the tires stick to the road you have control of the vehicle. Once the tires begin to slip, you lose control.

The dynamics of the interaction with the bow and the string can be seen in the following figure. The bow is moving upwards. The arrow indicates the location of the bow and the direction the bow is moving. The string moves with the bow during the stick phase, then rather quickly slips during the slip phase. The motion of the string at the bow location, corresponds exactly with the motion of the triangular wave that maps out the fundamental mode over one cycle. During the slip phase the string displacement changes quickly from upward to downward. Then during the rest of the cycle, at the location of the bow, the string is slowly moving upward with the bow. The triangular wave shape is complex and requires a superposition of many harmonics of the string.

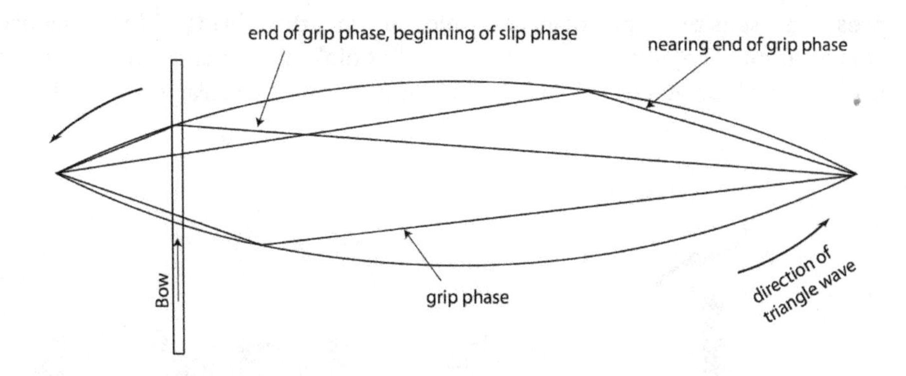

Instrument Body

Most stringed instruments, such as violin, guitar, and piano, have a similar acoustic construction. The string has a structural support called the **bridge** that holds the string above a vibrating plate, which is typically wood. A resonance between the harmonics of the string and the plate occurs. In a guitar, it is primarily the "top plate" that vibrates and produces the sound we hear. In a violin, both the top and bottom plate are important and are connected by a piece of wood called the "sound post." A piano has a "soundboard." This is what you see if you look at the back of an upright piano. By the way, you should check the condition of the soundboard if you buy a used piano. A piano with a slight crack in the soundboard may still sound quite good, but its market value is much lower.

Acoustic guitars have similar vibrational characteristics as violins, except that the back plate vibrations play a much less important role. We will discuss guitar top plate vibrations a little more in Chapter 10. There is a subtle but important conceptual issue here. When we talk about natural modes of the top plate, or any "plate," they are inharmonic – they do not form a harmonic series. In Chapter 10, we will go into more detail about what we mean by inharmonic natural modes. We are foreshadowing a bit.

We do need to mention the vibrations of the instrument body, however, to explain how stringed instruments work. As an interesting example, the Ovation Guitar Company takes advantage of the fact that the *guitar back plate vibrations are not very important* acoustically. The Ovation guitars use a plastic composite bowl, rather than a conventional wood back plate. The top plate, however, is traditionally manufactured of wood (spruce or cedar). The Ovation Viper guitar is shown below. Ovation guitars have a pressure sensitive piezoelectric pickup in the bridge for electric amplification. The guitar shown here has "f-hole" style sound holes which is fairly unusual for guitars. Most acoustic guitars have a circular sound hole.

Ovation Viper YM63 with a cedar (conventional wood) top, courtesy of Ovation Guitars.

Violin Body

Let's examine, in a bit more detail, the design of a violin body. The top plate is very thin wood (2-4 mm) and is coupled to the back plate with what is called the sound post. The bridge couples the vibrational energy from the strings to the top plate and pushes down quite hard (100 Newtons) on the top plate. A piece of wood called the bass bar is glued to the top plate to add structural support and mass to the top plate. The f-holes allow the air resonating inside the body to couple to the surrounding air and radiate sound energy.

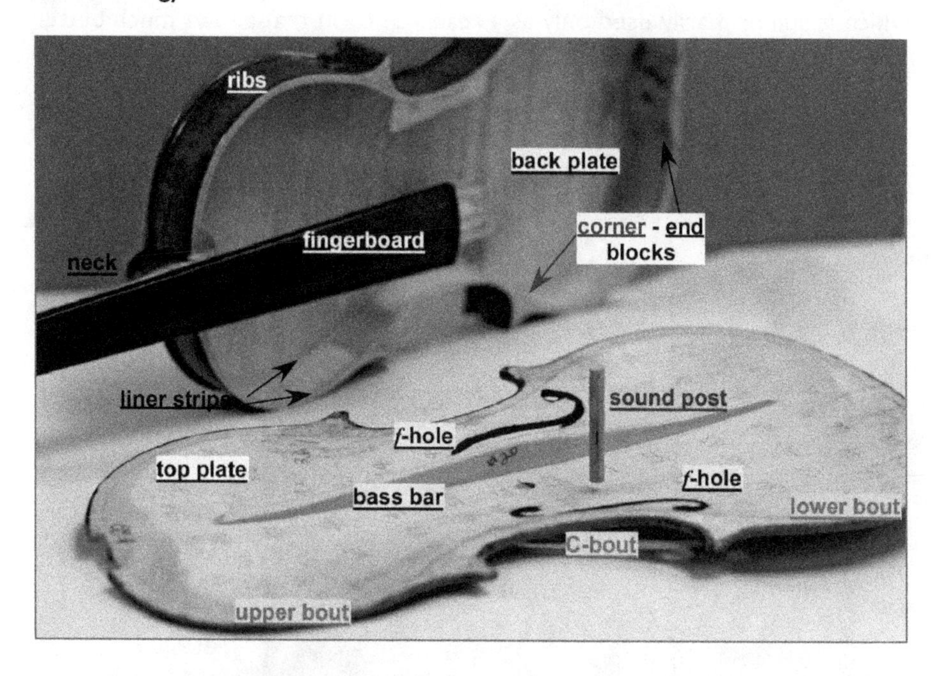

Parts of the violin body. Note that the sound post transmits vibrations from the top plate to the bottom plate. A guitar does not have a sound post. Graphic from G. Bissinger, East Carolina Univ.

The bridge plays a critical role in coupling energy from the string to the body. The bridge itself has resonances (the bridge is not shown in the figure above). It has different types of motion that resonate at different frequencies. It can rock back and forth in two directions. These resonances are at fairly high frequencies and can be lowered by adding mass to the bridge in the form of what is called a violin mute, leading to a darker timbre.

The natural modes of vibration – primarily of the top plate, back plate, and the air inside the violin – are what really radiate sound and determine what we hear. These can be studied by driving the body with a given frequency using a signal generator, amplifier, and speaker. One looks for resonances, frequencies where the body clearly vibrates at large amplitudes. The geometry or shape of the natural modes can be observed by putting sand on the surface and watching it jump around. The sand tends to settle at the nodes of the natural mode patterns. In Chapter 9, we'll discuss how some of these patterns can be seen using vibrating plates and sand.

There is a more sophisticated technique called hologram interferometry (which is still primarily used only as a research tool) that allows much better viewing of the patterns. Various patterns are shown below.

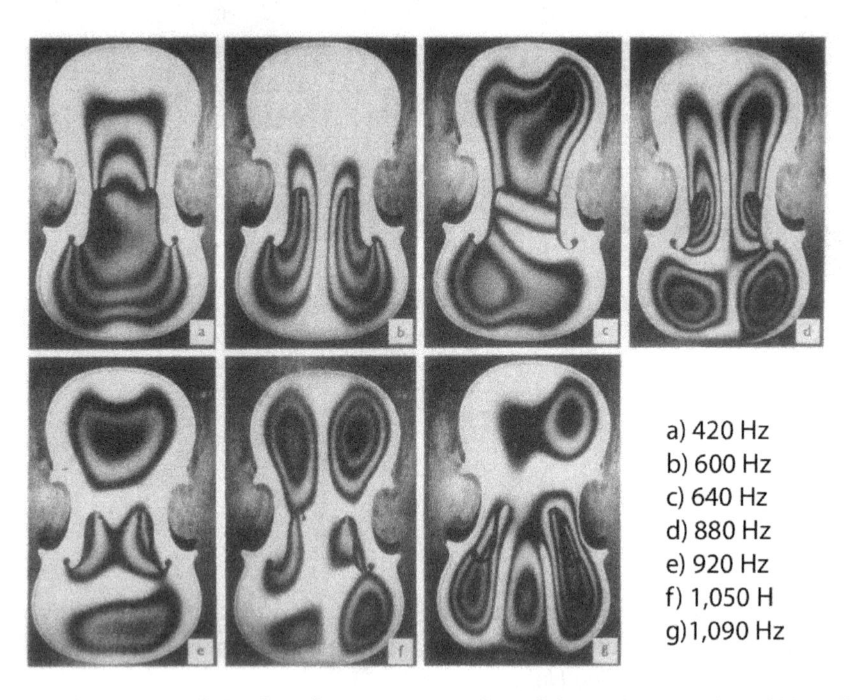

Vibrational patterns of the first few natural modes of the violin top plate. We will discuss the natural mode patterns in more detail in Chapter 10. From E. Jansson, et al. (1970)

In the study by Jansson, et al. that produced the images above, the top plate and bottom plate are isolated and vibrated individually. These vibrations will look somewhat different when the entire violin is intact, and the pieces all vibrate together.

These resonances do not have to be observed as visual patterns. For example, you could bathe the violin with sound from a speaker at specific frequencies using a function generator. A microphone attached to the body will detect increases in the resonance of the violin body (vibrations) at particular frequencies. You have to be a little careful about microphone placement, however, because if you are at a node on the body for a particular natural mode, you will miss that resonant frequency.

There is actually quite a bit of knowledge available about the violin, and there is what is called the "Catgut Acoustical Society," which is a group of scientists who publish books and have a scientific journal on the physics of the violin. A famous violinmaker Carleen Hutchins, who founded the society, designed a family of violins which have resonances scaled to match that of a fine violin. One of the design problems is that the geometry of these instruments does not necessarily correspond with a size that is comfortable to play. The difference in sound is noticeable and the viola equivalent sounds very good. One success of science in music!

References and Resources

Bissinger, G., "Parametric Plate-Bridge Dynamic Filter of Violin Radiativity," J. Acoust. Soc. Am. **132,** 465 (2012)

Catgut Acoustical Society Forum of the Violin Society of America See: http://www.catgutacoustical.org/

Fletcher, N. and T. Rossing, The Physics of Musical Instruments, 2nd Edition, 1998

Hutchins, C., "The Physics of Violins," Scientific American, November 1962

The Physics of Music: Readings from Scientific American, edited by C. Hutchins, 1974

Jansson, E., N. Molin, and H. Sundin, "Resonances of a Violin Body Studied by Hologram Interferometry and Acoustical Methods," Physica Scripta **2,** 243-256 (1970).

Rossing, T., F. Moore, and P. Wheeler, The Science of Sound, 3rd Edition, 2002

Schelleng, J., "The Physics of the Bowed String," Scientific American, January 1974

Chapter 9
Percussion and Inharmonic Vibrations

Chapter 9 Outline

- Inharmonic Natural Modes of Vibration
- Marimba, Xylophone, Vibes, and Glockenspiel
- Carillon Bells
- Timpani
- Guitar Top Plate

Natural Modes of Vibration

Throughout our discussion of vibrations and sound we have run across many examples of "things" that vibrate with frequencies that form a harmonic series, but also a few that do not. One good example of a thing that does not vibrate with natural mode frequencies that form a harmonic series is the top plate of a guitar or violin. It turns out that quite generally objects vibrate with natural modes that do not form a harmonic series and we call such frequencies "inharmonic." Do not be bothered with the term "mode," it is used in many different contexts, both in music and in physics. When we say "natural mode," we are referring to a particular vibrational pattern (the natural mode vibration) that happens at one particular frequency (the natural mode frequency).

Anything that springs back quickly will exhibit natural modes of vibration, and usually these mode frequencies are inharmonic. In this book so far, we focused on musical instruments that produce a distinct pitch, therefore they all had harmonic natural mode frequencies. Typically, percussion instruments do not produce the sensation of a distinct pitch and are inharmonic. There are many subtle exceptions – timpani, bells, etc. – that we need to address. We won't worry about the issue of categorizing the piano as a percussion instrument (an issue often raised). Suppose we drive a drumhead with a speaker with an amplified sine wave and vary the drive frequency slowly by turning a knob. We can see the shapes of the natural modes of vibration with the naked eye, or we can strobe the drumhead to get a slow motion effect to see the natural modes even better. These "mode" shapes are "regular" in some sense, but quite complex.

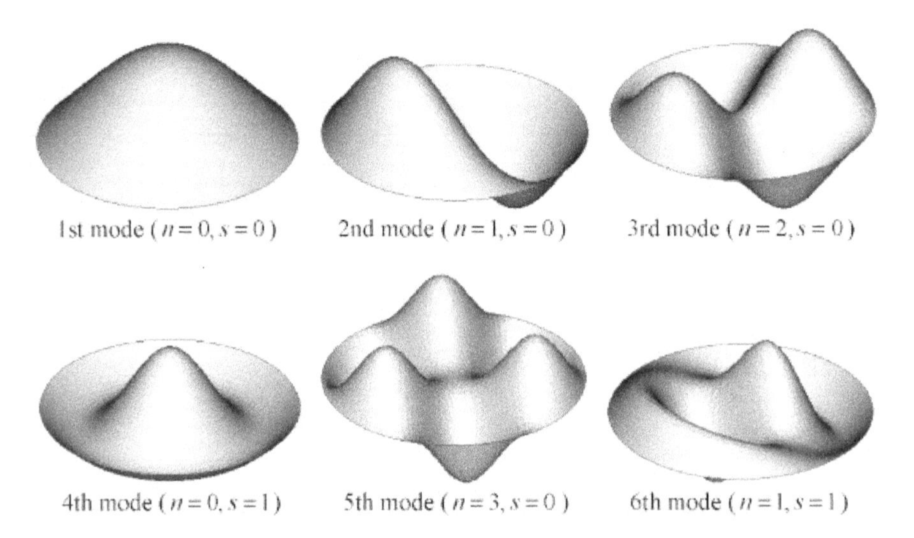

First six natural modes of a vibrating membrane from computer simulation. From W. Wu, C. Shu, and C. Wang, J. of Sound and Vibration, ***306*** *252-270 (2007).*

Only when we deal with long, narrow and uniform objects do we obtain a harmonic series for the natural mode frequencies. Why? Well, in these ideal situations, the object can sustain sinusoidal waves like the standing waves we have been talking about in Chapters 8 and 9. These objects can be (to a good approximation) treated as one-dimensional. Mathematical solutions to vibrations in one dimension produce sinusoidal solutions.

Another classic example from physics is the so-called "Chladni Plate" demonstration. Ernst Chladni (1756-1827) was a German physicist and musical instrument inventor. Like the drumhead example above, a metal plate can be vibrated using a mechanical vibrator. Sound waves alone are of too small amplitude for the demonstration to work. The mechanical vibrator can simply be a speaker coil attached to a rod that connects to the center of the plate. In the olden days, the person doing the demonstration (typically an instructor for introductory physics) would bow the side of the plate with a violin bow and generate resonant frequency vibrations between the bow stick-slip mechanism and the plate. See the image that follows.

Image of the various natural modes of a square Chladni Plate. Simpler patterns correspond to lower natural mode frequencies. Images originally from "Sound," John Tyndall (1869).

The Chladni Plate is painted black, and fine white sand is sprinkled uniformly on the surface of the plate. At certain frequencies (the natural mode frequencies of the plate), there is a resonance between the driver and the plate. The plate vibrates at large amplitude, and the fine sand moves around and off the plate. However, it is a little more complicated than that. There are node and antinode regions. In the node regions of these vibrational modes, there is little or no motion of the plate. The sand accumulates in these places. At the antinodes, there is a lot of motion and the sand is kicked around (mostly up and down, but sideways too) and moves to the node regions or simply just bounces off the plate. If the Chladni plate is vibrated the old-fashioned way with a bow, best results are obtained if the bow touches the plate at an anti-node location.

Chladni plates do not need to be square. They can be circular, oval, or even shaped like the top plate of a violin. In fact, a top plate of a violin behaves similarly to a Chladni Plate except for two things. First, the steel Chladni Plate is stiffer than the thin wood top plate. This, as you might expect, leads to higher natural mode frequencies. Second, the shape of a violin top plate is more complex. There are four key features of natural modes:

1) The complex vibrational mode patterns are determined by the shape and geometry of the object.
2) Each vibrational mode has with it an associated mode frequency.
3) The mode frequencies are typically inharmonic.
4) More complex vibrational mode patterns of a given object are associated with higher mode frequencies.

We will now discuss a few percussion instruments that have inharmonic mode frequencies (also called partials), but still produce a perceived pitch. These instruments have natural mode frequencies that line up in a particular way to be almost harmonic, or at least harmonic enough to fool the ear into hearing a distinct pitch. Such instruments include bars, such as marimba, xylophone, and vibes. Additionally, some bells produce a sensation of pitch, as do timpani (also called kettle drums). Many percussion instruments produce little sensation of pitch, such as a snare drum, a bass drum, or a gong. These instruments *still do have* natural modes of vibration and distinct natural mode frequencies. However, the frequencies are inharmonic.

Marimba, Xylophone, Vibes, and Glockenspiel

These instruments are bars struck by a mallet. The marimba and the xylophone are very similar, except the xylophone has a brighter timbre. The marimba is larger, has lower notes, and has a softer and darker timbre. The xylophone uses rubber-coated mallets, where the marimba uses yarn-wound coated mallets. Both instruments have resonator tubes below the bars, but the tubes are longer on a marimba. Both instruments use rose wood or synthetic wood bars. The vibraphone (or vibes) uses aluminum bars, which ring much longer. The vibraphone has mechanical motor driven "vibrator disks" that rotate within the resonator tubes adding vibrato to the sound. More details about these instruments can be found in Rossing (1992).

Five Octave Concert Marimba, courtesy of Adams Musical Instruments

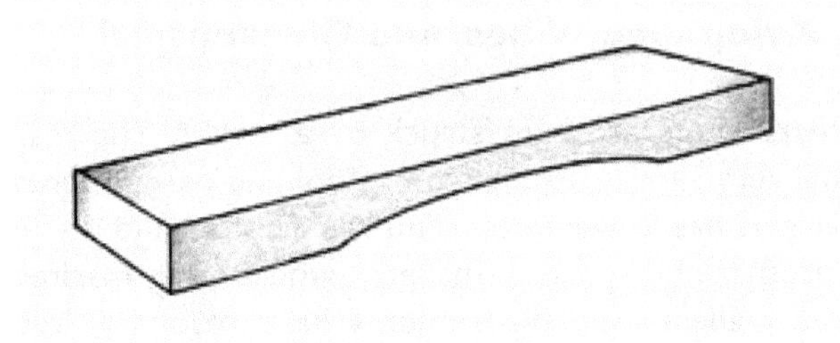

The marimba (wood), xylophone (wood), and vibraphone (aluminum) bars are concave on the bottom to enhance the natural mode vibrations. From: http://www.vsl.co.at/

The frequencies of a vibrating bar are inharmonic. However, the first natural mode can be tuned down by shaving a concave arch on the bottom of the bar. This reduces the restoring force of the first partial substantially and reduces its resonant frequency. This brings the frequency ratio of the first and second partial to 1:3 for a xylophone and 1:4 for a marimba and vibes. For orchestral bells (called glockenspiel, which are bars as well) the frequency range of the second partial (over 2000 Hz) is so high that it dies out quickly and does not impact the perception of pitch. The glockenspiel bars are not concave (carved out) on the bottom.

Carillon Bells

Many types of bells give the sensation of distinct pitch. Examples include carillon bells (controlled by a keyboard), hand bells, and orchestral chimes (hollow tubular bells). However, many other bells, for example a large church bell, a fire alarm bell, or a gong do not produce the sensation of a distinct pitch. Bells with a particular pitch are crafted in a particular way so that the natural modes (called partials) are "tuned" to produce the sensation of a particular pitch. We call the inharmonic natural mode frequencies of a musical instrument "partials." Different types of musical bells are tuned in different ways and these tunings are discussed in the Rossing (2002) text. Let's examine one particular example, namely, the carillon bell.

Carillon Bells from: Justin Ryan, http://www.carillontech.org

A carillon is a musical instrument (a large musical instrument) that consists of at least 23 bronze bells (two octaves). The carillon has a complex mechanical system to strike and damp the bells and is controlled by a keyboard. The carillon keyboard has a layout similar to a piano keyboard, but consists of pegs (called batons) that the player strikes with his or her fists. Foot pedals are used as well.

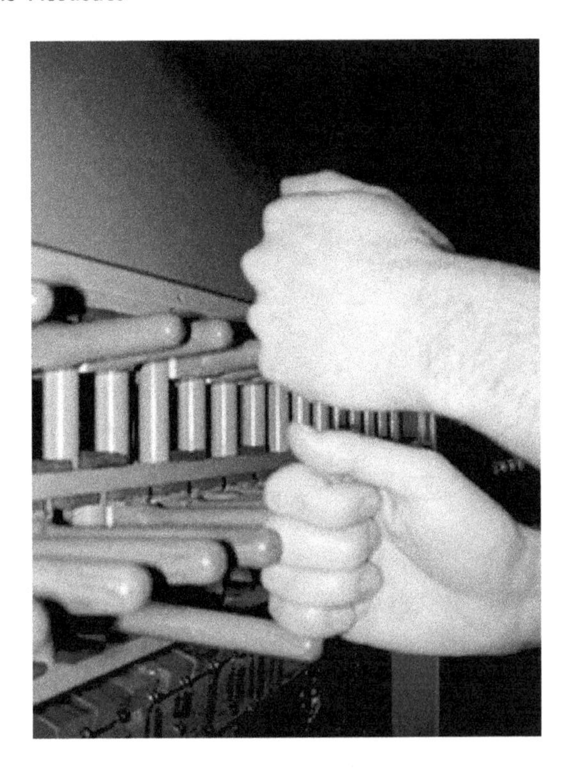

http://www.carillontech.org

The perceived pitch of a bell is called the **strike tone**. The carillon bell has the following natural mode frequencies:

0.5f , f, 1.2f, 1.5f, 2f, 2.5f, 3f, 4f, ... where f is the strike tone

The actual value of the strike tone (at frequency f) depends on the size of the bell. The lower frequency partial at 0.5 f is what is called the **hum tone**. Bells, in general, typically have a low frequency hum tone (though not always) at one-half the strike tone. A chime bell has a hum tone at 0.24 the strike tone. The natural mode shapes of the first five modes are shown below. The node regions are shown as dotted lines, and the antinode regions would follow towards the center of the regions outlined by the dotted lines.

The perceived pitch is determined primarily from the 2f, 3f, and 4f partials. This is known through studies where listeners hear only certain partials and are asked what the perceived pitch is. The partial at the frequency of f can actually be removed, and the (synthesized) pitch (and timbre to some extent) is not changed much. This perceived pitch based on the partials at 2f, 3f, and 4f is a good practical example of virtual pitch discussed at the end of Chapter 6.

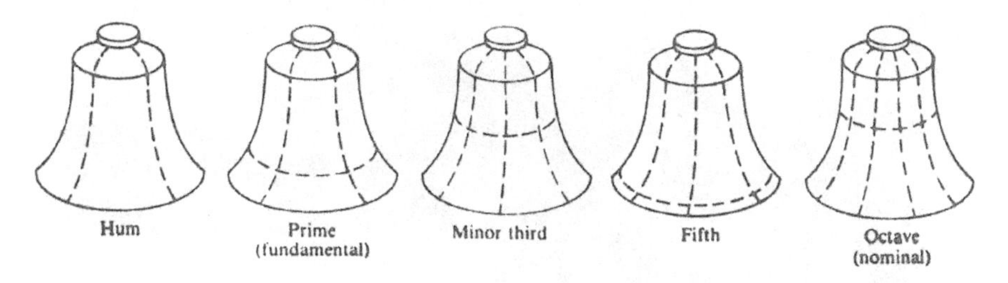

http://www.carillontech.org

Timpani

Timpani (also called kettle drums) are drums that produce a unique sound that is somewhat tonal. This is in contrast to a typical drum that sounds completely inharmonic. The timpani has a large drumhead (52 to 76 cm in diameter) held taut above a large copper bowl. The kettle or bowl can range from hemispheric to parabolic in shape and this helps control the tone quality. It turns out that timpani are the loudest of all orchestra instruments. Each timpani drum has a foot pedal that adjusts the tension on the drumhead allowing for the percussionist to tune the instrument. The timpani takes advantage of resonances between the drum head and natural modes shown in the first figure of this chapter and the air mass of the kettle to produce the a more harmonic sound.

Photo: Scott Albeman/Flickr

A circular membrane, like shown in the first figure of this chapter, has natural modes of vibration at inharmonic intervals. The frequencies of the first four modes are:

$$f,\ 1.59f,\ 2.13f,\ 2.29f\ \ldots$$

These frequencies are well determined, but require quite a bit of rather complex mathematics. These particular frequency ratios come from Table 13.3 of Rossing (2002). The mathematics involved is given in Fletcher (1998). The upshot is that this inharmonic series of frequencies that decay away rather quickly produces a percussive sound. This brings up an important point. Specifically, the decay rate of the partials has an enormous effect on our sense of timbre. For drums, in general, the decay of the partials is quick. This gives that percussive sound.

The timpani (or kettle drum) has a large enclosed air mass that is the right size and shape to align the partials in such a way to produce the distinct timbre and a sense of pitch. The measured frequencies of a kettle drum are:

$$0.85f,\ f,\ 1.51f,\ 1.99f,\ 2.44f,\ 2.89f\ \ldots$$

The perceived pitch called the strike tone, is at frequency f. The frequencies of all the natural modes are reduced due to the coupled air mass. Also, the fact that the drumhead is made out of Mylar or calf's skin and is somewhat stiff has an effect. Additionally, some of the natural modes are not present because of the location of the strike point of the mallet. The first natural mode is a global in-out oscillation, which couples well and radiates away quickly. The first natural mode turns out to be really just a "thump" because it dies away so quickly. The second natural mode is the prominent partial and is the frequency of the perceived pitch.

The fact that the timpani produces a somewhat vague sense of pitch provides an excellent example of the gray areas in musical acoustics. The basic concept that we need a harmonic series produces the sensation of a distinct pitch is not absolutely true! The examples of musical bars, bell, and timpani show the complexity of the perception of pitch and timbre, and the importance of virtual pitch.

Guitar Top Plate

In our discussion of natural modes of vibration, a fantastic example is the resonant "plates" used to acoustically amplify the sound of a string instrument. By "plates" we mean, for example, the top plate of a guitar, to a lesser degree the back plate of a guitar, the top and bottom plates of the violin, and the soundboard in a piano. We already mentioned the top and bottom plates of the violin near the end of Chapter 8. The plates in string instruments exhibit complex patterns like the Chladni plates. In fact the vibrational mode patterns are more complex because the plate shape is more complex.

As we mentioned in Chapter 8, the back plate on a guitar is less important than a violin. Its primary purpose is to produce an enclosed air mass which vibrates with the top plate. The spruce wood top plate couples to the string though the bridge and the key point is that there are resonances between the natural mode frequencies of the top plate and the harmonic frequencies of the string. This causes the top plate to vibrate, push and pull on the background air, and radiate sound waves. The resulting sound is much more intense than any sound waves that might be produced by the thin vibrating string alone. Let's examine what these natural modes of the top plate look like. The first 6 natural modes are shown on the following page.

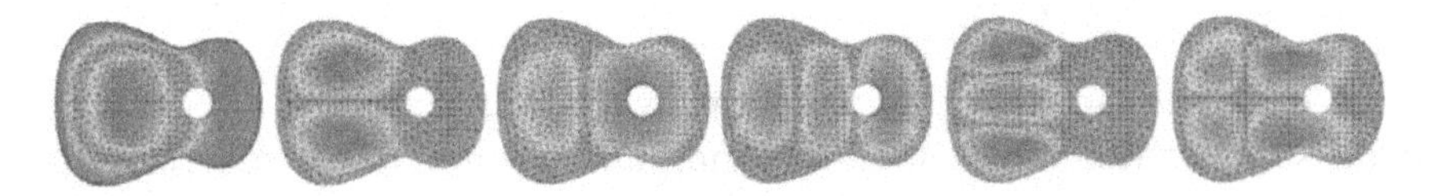

Numerical simulation results show natural modes of vibration shapes for a guitar top plate. From Fig. 9 of "Numerical Simulation of a Guitar," E. Becache, A. Chaigne, G. Derveaux, P. Joly, Computers and Structures, Vol. 83, pp. 107-126 (2005).

Researchers E. Becache, A. Chaigne, G. Derveaux and P. Joly (2005) studied the physics of the guitar using numerical modeling (finite element analysis) and successfully compared their results to the experimental measurements of E. Jannson (1971) using laser hologram interferometry.

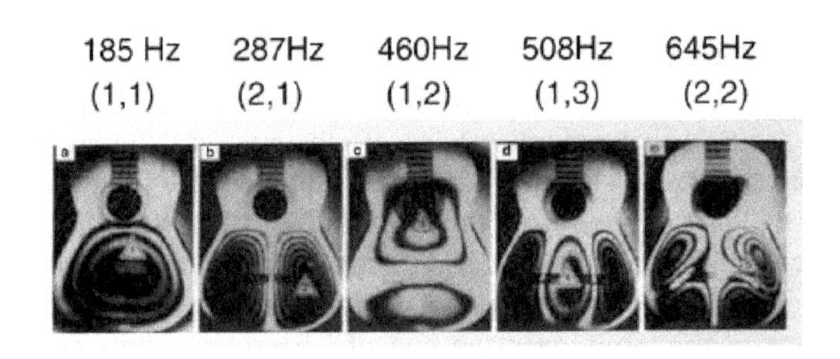

Actual measured natural mode vibrations using laser hologram interferometry. From Fig. 8 of "Numerical Simulation of a Guitar," E. Becache, A. Chaigne, G. Derveaux, P. Joly, Computers and Structures, Vol. 83, pp. 107-126 (2005).

References and Resources

Becache, E., A. Chaigne, G. Derveaux, P. Joly "Numerical Simulation of a Guitar," Computers and Structures, **83**, 107 (2005).

2009. See: http://carillontech.org also see: http://www.gcna.org

Fletcher, N. and T. Rossing, The Physics of Musical Instruments, 2nd Ed., 1998.

Hall, D., Musical Acoustics, 3rd Edition, 2002

Jannson, E., "A Study of the Acoustical and Hologram Interferometric Measurements on the Top Plate Vibrations of a Guitar, Acoustica, **25**, 95 (1971).

Rossing, T., F. Moore and P. Wheeler, The Science of Sound, 3rd Edition, 2002

Ryan, J., "Carillon Mechanics and Technique," Carillon News, Guild of Carillonneurs in North America, No. 81, April "Vienna Symphonic Academy - Vienna Academy" www.vsl.co.at -- This is a company in Vienna, Austria that produces high quality digital samples of musical instruments. The "Vienna Academy" link provides a detailed description of all the different symphonic instruments.

Chapter 10 Woodwind and Brass Instruments

Chapter 10 Outline

- Standing Waves in a Pipe Open at Both Ends
- Standing Waves in a Pipe Closed at One End
- Conical Bore Instruments
- Woodwinds with Reeds
- Tone Holes
- The Register Key and Octave Key
- Brass Instruments
- The Flared Bell
- Flute and Other Edge-Tone Instruments

Standing waves in a Pipe Open at Both Ends

All brass and woodwind instruments have long tube-like resonant cavities. Let's look at a few very simple musical "pipe systems." We will later see that these pipe systems are excellent simple models (and sometimes fairly accurate simple models) of a variety of instruments. The first musical pipe system we will examine is a long and narrow pipe open at both ends. The pipe open at both ends actually turns out to be a very good approximation of a flute. If the pipe has too large a diameter the simple concepts we describe here must take into account the radial behavior of the air column and not simply the wave motion along the length of the tube. Do not worry too much about these corrections at this point. We had a similar issue when we assumed our strings had no stiffness. The standing waves (or natural modes) in a pipe open at both ends look familiar to us and are analogous to the natural modes on a string.

For a pipe open at both ends, we actually find standing sound wave patterns. These standing waves "fit in the pipe" and naturally resonate to fairly large amplitudes. What do we mean by "fit in the pipe?" Because the pipe ends are open, the pressure is fixed at atmospheric pressure, and the amplitude of the pressure fluctuation is forced to zero. Compressions and rarefactions can "build up" inside the pipe where the air is confined by the tube walls.

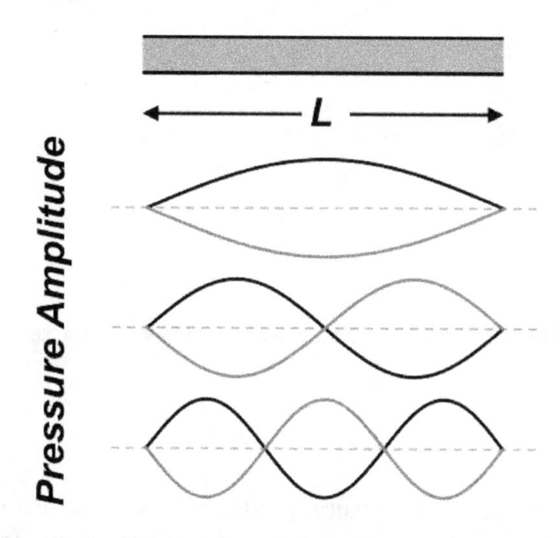

Displacement is driven by a spatial *change* (or gradient) in pressure, so where the pressure is changing the most, such as at the ends, the amplitude of the displacement is at a maximum. Where the pressure is changing the least, at the maxima and minima of pressure, the amplitude of the displacement is forced to zero.

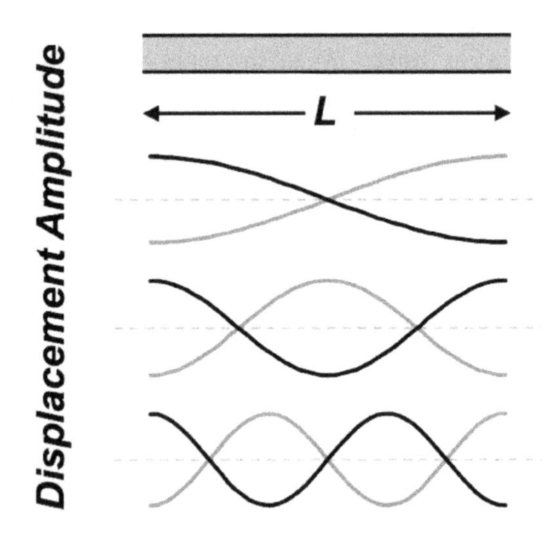

The wavelengths of the natural modes follow the pattern below.

$$\lambda_n = 2\,L / n$$
$$(\text{for } n = 1, 2, 3, \ldots)$$

The frequencies of the natural modes follow a pattern similar to those of a string EXCEPT the wave speed is the speed of sound and NOT the speed of a wave on a string (involving string tension, mass, and length). The speed of sound in air is about 344 m/s. For the natural modes of the pipe open at both ends, the frequencies follow the pattern below.

$$f_n = v_s / \lambda_n$$
$$(\text{for } n = 1, 2, 3, \ldots)$$

$$f_n = n\,v_s / (2\,L)$$

$$f_n = n\,f_1$$

Standing Waves in a Pipe Closed at One End

Standing waves in a pipe closed at one end are just a little different from those within a pipe with two open ends. The clarinet and related instruments are represented pretty well as a pipe closed at one end. At the closed end, the displacement of the air must go to zero because the air bumps up against a "wall." However, at the closed end, the pressure can build up. The open end behaves the same as the open ends for the pipe open at both ends.

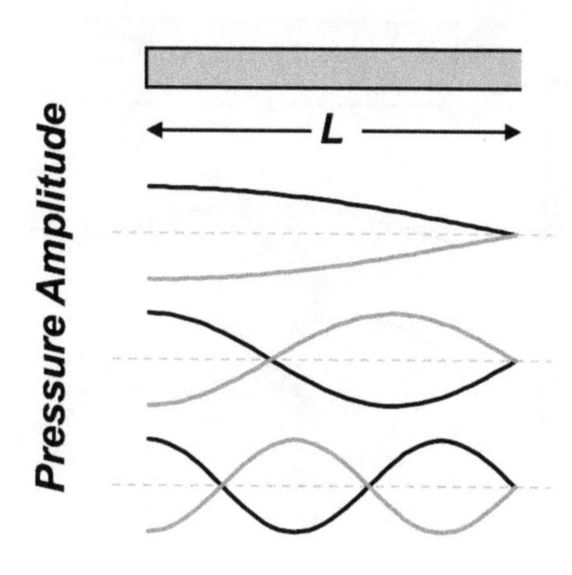

The most interesting result of closing one end of a pipe is that only odd harmonics are present for a pipe closed at one end. The even harmonics we are familiar with for both the string and the pipe open at both ends do not exist!

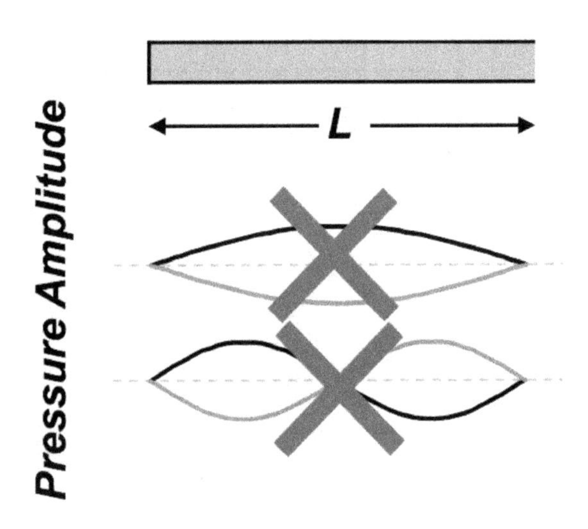

The wavelengths of the natural modes for a pipe closed at one end follow the pattern:

$$\lambda_n = 4\ L\ /\ n, \quad (\text{for } n = 1, 3, 5, \ldots)$$

The frequencies of these natural modes follow the pattern:

$$f_n = n\ v_s\ /\ (\ 4\ L\ ), \quad (\text{for } n = 1, 3, 5, \ldots)$$

$$f_n = n\ f_1, \quad (\text{for } n = 1, 3, 5, \ldots)$$

The clarinet has a cylindrical bore up to the bell and behaves like a pipe closed at one end. On a clarinet, the sudden flared bell increases the radiation of sound out of the instrument. You can actually play a clarinet without the flared bell (it detaches), and it works pretty well. The fact that the clarinet produces only odd harmonics gives the instrument its unique "woody" sound, even if the clarinet is made of plastic.

Conical Bore Instruments

For a conical bore, the pressure fluctuation is at a maximum at the narrow closed end and at a minimum at the open end. It turns out that even though one might think the conical bore behaves like a pipe closed at one end, quite the opposite is true! For the conical bore chamber closed at one end, the natural mode frequencies form a complete harmonic series. That is, the frequencies are integer multiples of the fundamental. This result is not at all obvious, but can be shown by solving the "wave equation" in a conical tube. In addition, the frequency spectrum can be measured to confirm that a conical tube closed at one end does indeed produce all harmonics. For more detail about the conical cavity and its natural modes, see the book "The Physics of Musical Instruments," by Fletcher and Rossing (1998) that is referenced below.

Instruments, such as the saxophone and the oboe, have a conical bore that causes these instruments to have all integer harmonics present in their sound spectrum. The larger conical bore on a saxophone is a more efficient radiator, making the saxophone a much louder instrument than either the oboe or clarinet. Regarding the effect on timbre, the clarinet has a unique woody or hollow sound due to the missing harmonics. Clarinetists, please do not take offense to these comments coming from a sax player!

The oboe and the saxophone have a sound that is a little more cutting, and can be heard well, due to the fact that all harmonics are present. On the other hand, the clarinet has the advantage that higher natural mode frequencies go up by a factor of 3, 5, 7, ... from the fundamental. This allows higher registers to be played by the clarinet. Thus, the clarinet has a wider range of pitch than the saxophone or oboe. So, if being heard is an asset, the clarinet ultimately becomes the winner, because of the very high notes it can play! Of course, nobody can compete with the trumpets for volume. There is a guy on YouTube who breaks a wine glass playing his trumpet without any amplification. You have to check it out.

http://www.youtube.com/watch?v=Pq-PxdOarjA
(or search: Shattering Glass with the Sound of a Trumpet!)

Woodwinds with Reeds

The reed vibrates in resonance with the natural modes of the attached tube. The vibrating reed creates pressure pulses in sync with the natural mode pressure fluctuations within the tube. When the reed swings open the pressure is at a maximum and the source of the pressure is the instrumentalist blowing into the mouthpiece. Then, the reed swings closed and the pressure is at a minimum, again in resonance with the natural modes in the tube. If you disconnect the mouthpiece, it vibrates at a quite high frequency, and the frequency is sensitive to embouchure pressure on the reed. (Embouchure is the posture of your lips and mouth.) When you connect the mouthpiece to the instrument, the reed syncs up with the natural modes within the tube, and it is very clear there is feedback between the resonant cavity and the reed vibrations.

A good exercise for a reed player is to play only the mouthpiece unattached from the instrument body and maintain a steady tone. This is very hard to do without the resonance of the body "locking in" the particular frequency/pitch.

Tone Holes

In woodwind instruments, there are tone holes (also called key holes or side holes) that effectively truncate the length of the instrument at the location of the hole (if they are large enough). Smaller holes have a less dramatic effect and the effective length of the instrument is a little longer than the distance to the tone hole. For a large tone hole, we simply take the length of the tube from the source end to the open hole to be L in the various formulas for natural mode frequencies used throughout this chapter.

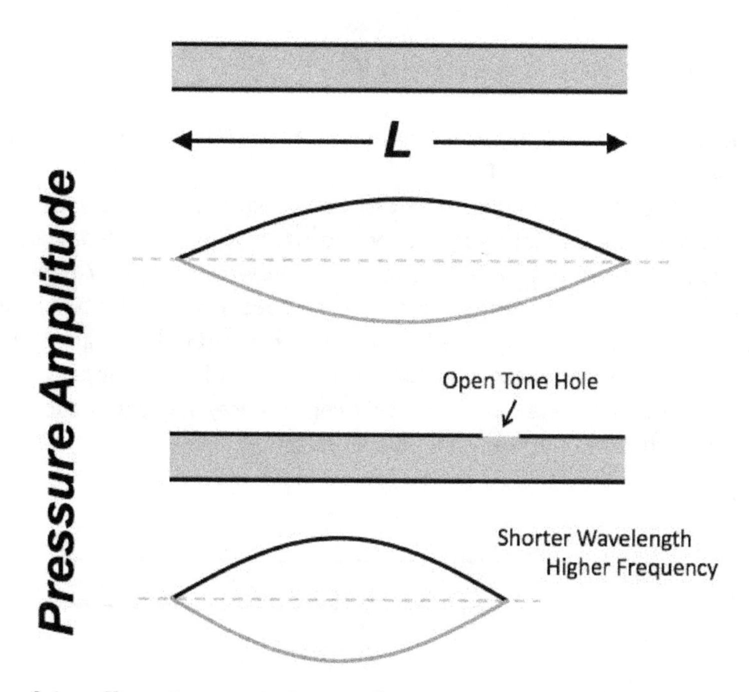

Example of the effect of a tone hole on a flute-type instrument (pipe open at both ends)

Example:

What is the frequency of the fundamental of a 1 m long flute with the first open tone hole located at 0.2 m from the end furthest from the mouthpiece. Assume the tone hole is large.

Solution:

f_1 = 344 m/s / (2 L) (Frequency of the fundamental)

L = 1 m - 0.2 m = 0.8 m (L is now the length to the open side hole.)

f_1 = 344 m/s / (2 x 0.8 m) = 215 Hz

The Register Key and Octave Key

What happens when you press a register key on a clarinet or an octave key on a saxophone?

With a reed instrument, one can accentuate higher harmonics with a tight embouchure and blowing fast while narrowing the throat and mouth cavity. This method is not the easiest way, however, to play high notes. A small hole placed at the pressure antinode does a better job. The pressure maximum cannot be supported with the now-open hole, and harmonics with pressure anti-nodes at the location of the hole are suppressed. In the cylindrical bored clarinet, pressing this register key increases the frequency by factor of three (a perfect fifth plus an octave).

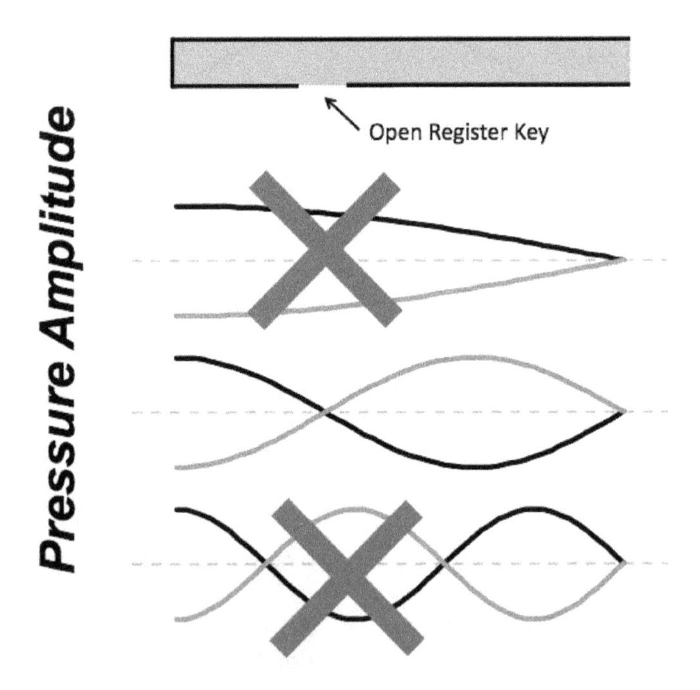

In a conical bored saxophone or oboe, use of the octave key increases the frequency by a factor of 2. The octave change makes the fingerings of the notes the same. On a flute, higher octaves are achieved primarily by what is called "over blowing" (increasing the velocity of the air stream from the mouth/lips) and using alternate fingerings that accentuate the higher harmonics in a manner like an octave key on a saxophone.

One issue that arises with octave keys for saxophone and oboe is that the location of the octave key shifts when different notes are played. A saxophone has two octave keyhole locations, and one or the other is used depending on the note being played. An oboe has three octave keyhole locations. A clarinet has only one register keyhole. A compromise is made in the design so that all notes sound favorably when the register/octave keyhole is open. Remember the instrumentalist has additional control of which octave is being played by controlling airflow and embouchure.

Brass Instruments

For brass instruments, the buzzing lips play the role of the reed. The lips are more massive than a reed, and the instrumentalist has a little more control over what frequencies are excited. This can be shown by comparing what a brass player can do with his mouthpiece versus the control a woodwind player has over the unattached mouthpiece. When the lips separate, a puff of air or a pressure maximum is sent into the horn and when the lips close, there is a pressure minimum. It is the feedback of these oscillations with the pressure antinode that allows the horn player to play particular tones.

The brass mouthpiece and lips, like the reed, behave acoustically like a pipe closed at one end. A horn without a bell will have overtones which are odd integer multiples of the fundamental like a pipe closed at one end.

The Flared Bell

As we mentioned above, the sudden flare in the bell of a clarinet is not acoustically significant except to increase volume in the lower range of the instrument. For brass instruments, the situation is completely different! If a trumpet did not have a flared bell it would behave as a pipe closed at one end and would have only odd integer overtones. The flared bell (and to a lesser extent the mouthpiece) cause the overtones to be harmonic (almost). This is due to many years of artisans tinkering with various shapes and sizes of the instrument bell. The original first natural mode is shifted upward and is not related in a harmonic way with the other normal modes, which take on a frequency of 2f, 3f, 4f, ... with the fundamental missing. The first natural mode can be played by careful control of the lips and is called the "pedal tone".

The natural modes of flared bells in combination with cylindrical and conical cavities are discussed in "The Physics of Musical Instruments", Fletcher and Rossing (1998). If you are interested in the actual solutions to the wave equation in these more complicated geometries, this is a good place to start.

The French horn is hard to play because many of the notes are excitations of very high harmonics (or overtones), which are relatively weak resonances and require enormous control of the embouchure. The French horn has quite a long acoustical length (375 cm), and the notes in the mid-range of the instrument involve exciting the higher harmonics of the instrument.

The flared bell is also important for increasing the radiated power (as it is for the clarinet). The bell allows for efficient coupling of the energy of rather large pressure fluctuations inside the instrument to the small pressure fluctuations of sound waves leaving the instrument. The bell is more efficient at radiating high frequencies. Also, from what we know about diffraction, high frequencies will be much more directional when coming from the bell.

The primary effect of the flared bell is to bring the overtones into harmonic relation. That is, 2f, 3f, 4f, 5f, ... The "fundamental" is shifted too, but is not an integer multiple of the overtones so is not of much musical use. We also said that the bell is more efficient at radiating higher frequencies, and this causes the higher harmonics to be less pronounced. Why? Because the overtones radiate away, they cannot feedback energy to the mouthpiece/lips to be enhanced by positive feedback between the sound source and the resonant cavity. Why does a horn shape enhance the efficiency of radiating the higher frequencies? This is a difficult question to answer. The fact that a horn is a more efficient radiator of sound waves makes sense because the resonant pressure fluctuations are coupled to a much larger volume of air in the bell. Lower frequencies are less directional and the amount of power radiated is less sensitive to the shape of the radiator (unless it becomes comparable to the wavelength).

Flute and other Edge-Tone Instruments

A jet of air is unstable and wants to mix and equilibrate with the background air. As a simple example, suppose you blow air through a straw. The air jet flowing out of the straw will be unstable and will begin to wobble and form a wavelike structure. Eventually, swirls are generated, then turbulence, then farther away from the end of the straw (beginning of the jet stream) there is little left to tell there was a jet stream to begin with. These wobbles of the air stream are the origin of flute-type sound sources or so-called "edge tones."

The figure on the following page shows a stream of smoke (which is a stream of air with visible particles in it). One can see how the smoke stream wobbles and generates swirls, known as vortices in the scientific study of fluid motion.

Photograph of a smoke stream. You cannot see air wobble, but if you add some smoke particles, you get some idea of what is going on. Photograph by Alex Stepanov

The next figure below is a numerical fluid dynamics simulation of edge tone flow of air. The swirls are known as "von Karman vortices." Theodore von Karman (1881-1963), a Cal Tech scientist studied a similar process of vortices produced by airflow around a cylinder.

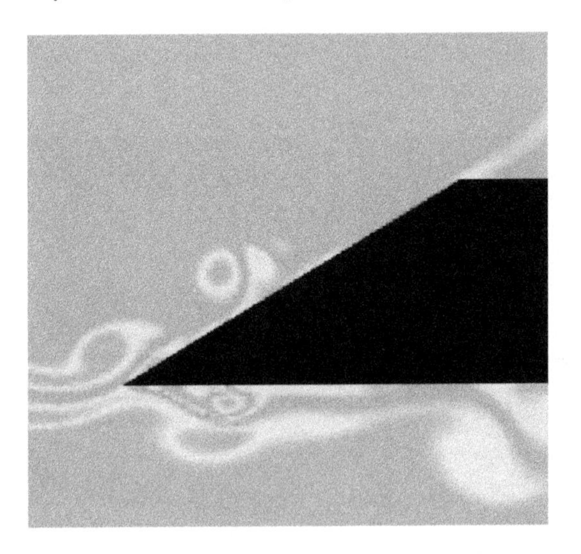

Fluid simulation of an edge tone by Lukasz Panek

In summary, sound is produced in a flute-type instrument by an air stream blown at a sharp edge, called an edge tone. The air stream oscillates above and below the edge. Sound sources for all flutes, flue organ pipes, recorders, and whistles, all work using this same principle.

Like with the reed (and brass instruments), the oscillating air stream couples with the resonant modes in the tube. Unlike the reed, the edge tone does not create pressure minimums and maximums but rather flow minimums and maximums. When the air jet flows into the pipe, the flow is at a maximum or inward; when the flow goes outside the pipe, the flow in the pipe is at a minimum or outward. Therefore, the edge tone behaves as an open end of a pipe. In reality, there is higher pressure near the edge tone source and tuning is necessary to make the flute harmonic. This tuning is implemented by tapering the pipe slightly near the embouchure hole. This part of the flute is called the head joint. The head joint is a very important part of the flute. Sometimes the head joint will be made of solid silver or even gold, whereas the body is manufactured with less expensive material.

To do a good first approximation, you can estimate frequencies of the harmonics of a flute by assuming it is a pipe open at both ends.

$$\lambda_n = 2\,L / n$$
$$(\text{for } n = 1, 2, 3, \ldots)$$

$$f_n = n\, v_s / (2\,L)$$

$$f_n = n\, f_1$$

Of course, the discussion of bore holes/keys and octave holes/keys still applies. The basic physics of flute-type instruments, specifically, the edge tone and the fact that the system behaves like a pipe open at both ends, applies to other instruments as well. This "flute-type" instrument category applies to recorders, some whistles, and some organ pipes as well.

Finally, one last comment on natural modes of wind and brass instruments... a very good exercise for improving tone for flute and saxophone players is to play the lower notes of the horn, say low B^b for the saxophone or flute. Then play the higher harmonics, by embouchure and throat (back of tongue) position. For saxophone, throat and tongue control should be emphasized. Intermediate players are surprised how many notes that can be played with one finger position in this manner. This exercise really gives you control of the harmonics of your instrument. Sigurd M. Rascher has a famous book "Top Tones for the Saxophone" (1994) describing these exercises as a way to play high notes outside the conventional range of the instrument, the so-called altissimo range. For brass players, there is absolutely no mystique to playing the harmonic series with one fingering. The bugle has no valves or buttons. Brass players learn about playing higher harmonics from day one.

References and Resources

Benade, A., "The Physics of Wood Winds," Scientific American, October (1960)

Benade, A., "The Physics of Brasses," Scientific American, July (1973)

Fletcher, N. and T. Rossing, The Physics of Musical Instruments, 2nd Ed., 1998.

Hall, D., Musical Acoustics, 3rd Edition, 2002

Rascher, S., Top Tones for the Saxophone, (1994)

Rossing, T., F. Moore, and P. Wheeler, The Science of Sound, 3rd Edition, 2002

Chapter 11
The Human Voice

Chapter 11 Outline

- Anatomy of the Vocal Tract
  - Vocal Folds
  - Vocal Tract
- Musical Formants
- Singer's Formant
- Raised Formants
- Throat Singing

The human voice is a truly remarkable musical instrument. It has the capability of drastically altering timbre, or the quality of sound, of a steady tone. The voice is much more versatile than the other acoustical instruments. Electronic synthesizers can widely vary the timbre of a tone, but with nowhere near the control of the human voice. The voice has an enormous variety of sounds and articulations. These different types of sounds are called "phonemes." The human voice allows human beings to communicate in ways well beyond that of other animals. Anthropologists have hypothesized that this is what sets the human being apart from his predecessors in the evolutionary process. At some point, the human vocal tract developed to a point to where it could communicate efficiently. If you look at the anatomy of other animals, they do not have the long vocal tract humans have. In fact, most mammals have the capability to eat and breathe at the same time, which is a big advantage for survival. They can truly "eat and run." Adult humans do not have this capacity because their larynx is deep and open to provide a resonant cavity for making vowel sounds. A diagram of the human vocal tract is shown in the figure below. We will discuss the anatomy of the vocal tract a little more. The two figures that follow come from the famous textbook called "Gray's Anatomy" written by Henry Gray and originally published in England in 1858.

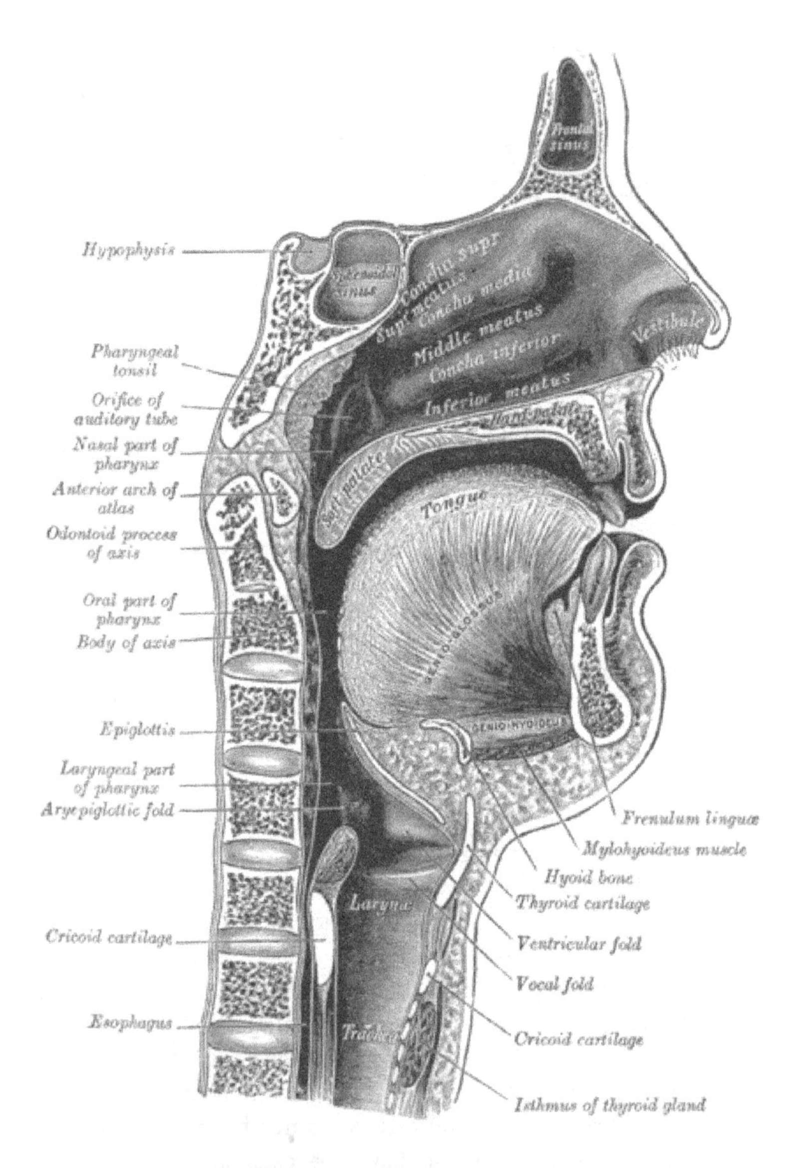

From "Gray's Anatomy," Henry Gray (1918)

Vocal Tract

The vocal tract is an open cavity including the nasal cavity, mouth cavity, pharynx, and larynx. The pharynx is the throat region behind the back of the tongue and above the larynx. The larynx is the "voice box" region and includes the vocal folds. The vocal tract (the acoustically relevant part of the mouth, nasal cavity and throat) ends at the vocal folds, which are made of soft skin-like tissue or mucous membrane material. The vocal folds are often called vocal chords.

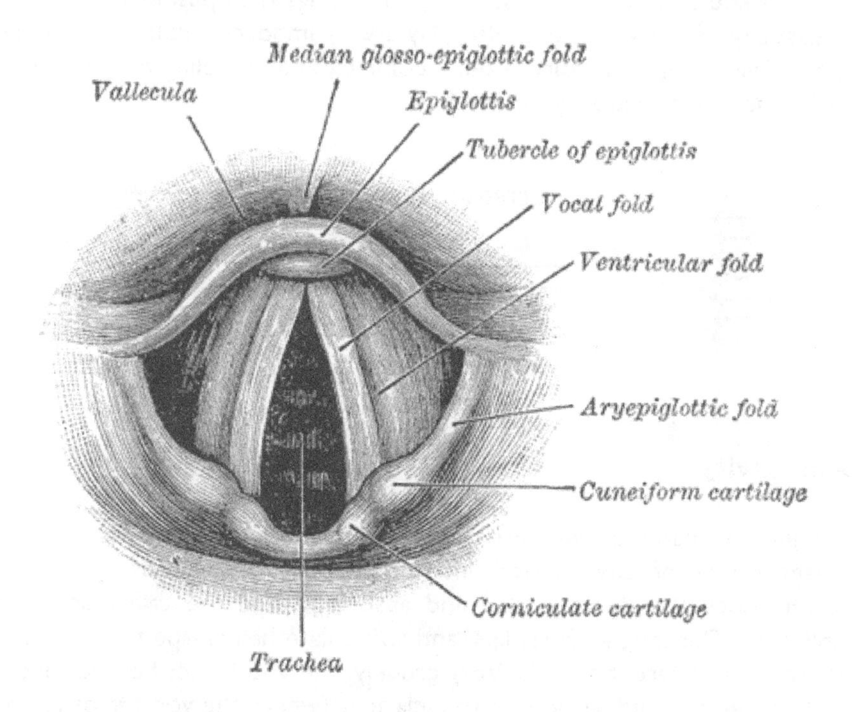

From "Gray's Anatomy," Henry Gray (1918)

The voice is like other wind instruments in many ways. There is a sound generating device and a resonant cavity. Air is forced from the lungs up through the trachea. The air stream passes through the vocal cords, which vibrate. The sound generation mechanism for the human voice is very similar to a brass player's lips buzzing in a mouthpiece. The vocal folds are about 2 cm long. Frequency depends on the tension supplied by small muscles attached to the vocal folds and supporting cartilage. The frequency range of vocal cord vibrations is approximately 70-200 Hz for men and approximately 140-400 Hz for women. There is an important physical effect called the **Bernoulli Effect,** which says that when a fluid speeds up, the pressure drops. This makes sense since if a small volume of fluid increases in speed, something must be pushing on it; so there must be a force on the fluid element, hence a pressure drop. We caution the reader that the Bernoulli Effect is only one aspect for understanding the vocal tract. As the constricted air flows faster through the folds, there is a pressure drop and hence a force pulling the folds together. When, the folds come together, the flow is blocked and the air pressure from the lungs then pushes them apart. The elasticity of the vocal folds also play a very important role. The actual motion of vibrating vocal folds is quite complex and modeling the vocal fold vibrations is an active area of research.

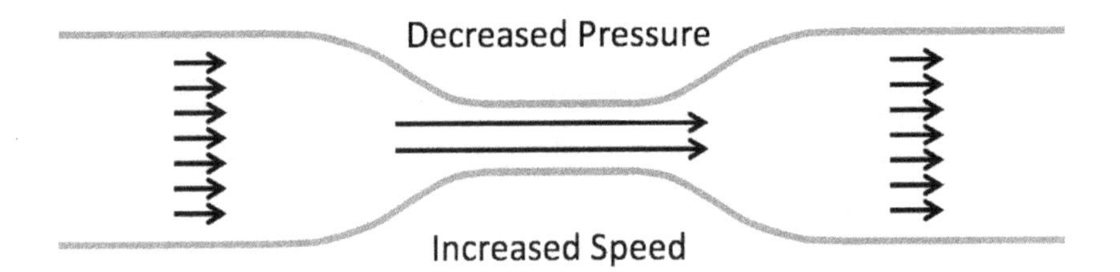

Vocal Cavity

Sound resonates in the throat cavity and mouth cavity, just like the resonant cavity of any musical instrument. The resonances are weak however, since the tissue is soft and absorbing, and the cavity shape is complicated. The tongue, jaws, lips, and soft palate help shape the resonant cavity to form different sounds. Very crudely, the vocal tract behaves like a pipe closed at one end, analogous to a clarinet (where the vocal folds are at the closed end, just as the reed is at the closed end of a clarinet). There is a huge difference between the voice and the clarinet, in that the resonant cavity is less selective than the long narrow pipe of the clarinet. This causes the perceived pitch to be determined by the vocal cords themselves. The buzz of the vocal cords would have the same vibrational frequency with or without the vocal tract acting as a resonant cavity (this is different than a clarinet mouthpiece). The vocal cavity selectively shapes the frequency spectrum. Certain harmonics are accentuated, and others are reduced. The importance of the shape of the vocal tract can be demonstrated using a buzzy duck call attached to different resonant tubes each having different shapes. There is a demonstration of this at the Exploratorium Museum of Science in San Francisco. Check out:

http://www.exploratorium.edu/exhibits/vocal_vowels/vocal_vowels.html

The "duck call" is shown below, which is a vibrating reed connected to a flowing air source.

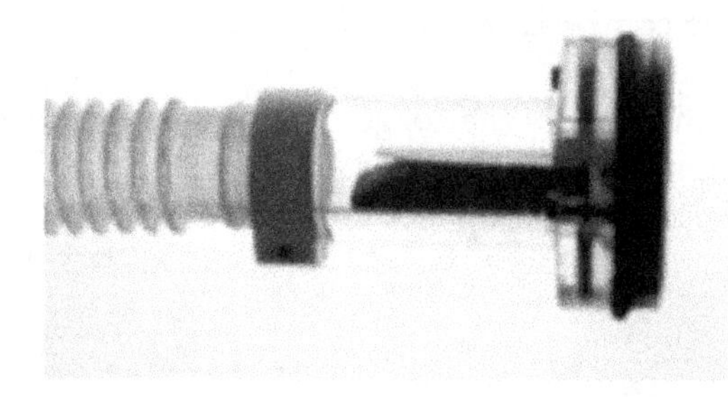

The duck call is attached to one end of various resonant chambers, two of which are shown below. The top chamber produces the "EE" sound, and the bottom one produces the "AH" sound.

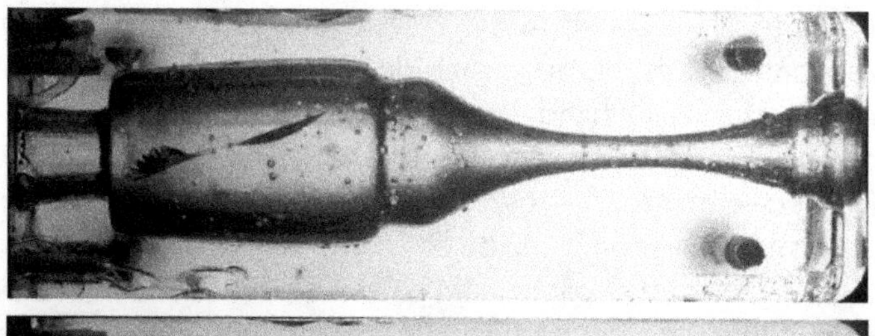

"EE" sound tube

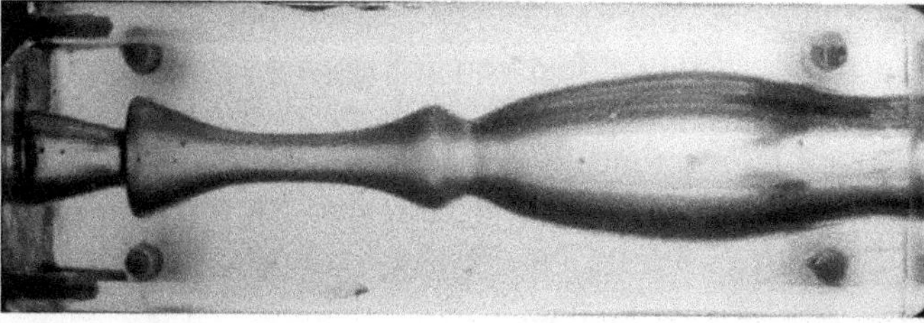

"AH" sound tube

http://www.exploratorium.edu/exhibits/vocal_vowels/vocal_vowels.html

A quite amazing aspect of the human voice is its ability make different sounds so rapidly. We don't realize how much mechanical motion is involved in shaping our vocal tract when we speak since we do so unconsciously. It's well worth watching a video of the vocal tract in action. Search "MRI Speech" on YouTube and take a look at a few of the magnetic resonance imaging (MRI) images of humans speaking or singing. One video of Joy Nash produced at the University of Southern California, is especially striking:

http://www.youtube.com/watch?v=0-aEN2xHBCc

The motion of the back of the tongue during speech is truly remarkable. We are not so surprised by the motion of the lips, mouth, and front parts of the tongue during the pronunciation of consonants (called plosive phonemes in phonetics) because we are conscious of this motion, and it is visible to us.

Formants

There are many types of sounds or phonemes generated when we speak. You may have heard of the categorization of various consonant sounds studied in speech and hearing science as well as in the field of linguistics. For example, consonant sounds can be categorized as plosive sound (e.g. "p") or a fricative sound (e.g. "f") that may be held. We will not cover these consonant sounds, because they do not generate the sensation of a distinct pitch. On the other hand, vowel sounds are steady sounds with a definite pitch. For example: *ah, ee, oo.* These vowel sounds produce complex periodic waveforms. A vowel held steady, has a frequency spectrum, which forms a harmonic series. It is the strength of the various harmonics that determines the particular vowel sound. Different vowel sounds are somewhat like different timbres. An experiment you can try is to hold a vowel sound steady. You will find you are singing a note. In fact, you cannot help but sing a distinct pitch when you hold a vowel sound steady. The pitch is determined by the fundamental frequency of vibration of your vocal folds. The specific vowel sound is determined by the shape of the frequency spectrum which is controlled by the resonances of your vocal cavity. We call these resonances "formants."

The **formants** are the natural modes or resonances of the vocal tract. The vocal tract can be crudely approximated as a pipe closed at one end.

The vocal cords are the source and cause pressure maxima and minima at one end, like the mouthpiece on a wind instrument. This means that the frequency location of the peak of the formants are at f, 3f, 5f, ... where f is the fundamental frequency of the "pipe" (the vocal tract really) closed at one end. For example, if the first formant is at 500 Hz, then the second formant is at 1500 Hz, and the third formant is at 2500 Hz. This corresponds to a vocal tract of 17.5 cm and approximates the location of the formants of the "ea" sound in "head," that has formants at 1500, 1800 and 2500 Hz. It is typically the first two formants that are important for recognizing vowel sounds. However, trained singers have 4 or more distinct formants. You will not see the higher formants for an untrained singer. You will sometimes hear singers who may have a pleasant warm voice, but it will be panned as "weak" due to the lack of the presence of the 4th and 5th formants.

These resonances, which we call formants, are quite broad. The vocal tract is made up of soft walls and has a complicated geometry. The original sound source is the vocal chords that generate the underlying harmonic series of frequencies. The formants (resonances) enhance the harmonics in a given frequency range. The shaping of the frequency spectrum by the complex shape and structure of the vocal cavity is a mechanical analog to what is know in electronic music as subtractive synthesis. It is also similar to a clarinet in some ways, in that it is approximately a pipe closed at one end. However, it is very different than a clarinet, in that the vocal tract resonances are broad, or not very distinct. In a clarinet, the body resonances are so strong and narrow in frequency that the reed vibrates only at the resonant frequencies of the body (or, a pipe closed at one end).

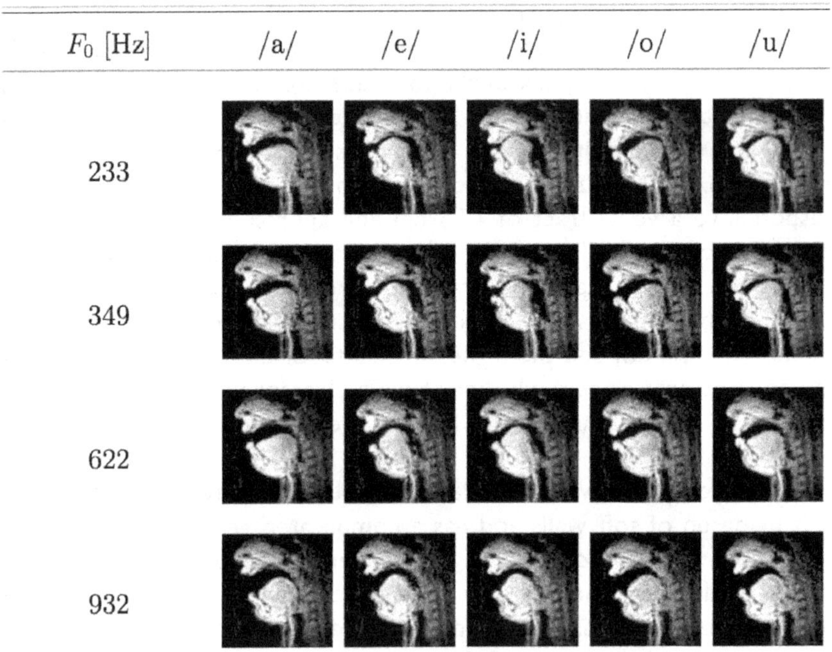

F_0 [Hz]	/a/	/e/	/i/	/o/	/u/
233					
349					
622					
932					

From E. Bresch and S. Narayanan (2010). Shown is the vocal tract of a soprano singer singing 4 different notes with the fundamental frequency pitch given as F_0. Different vowel sounds are shown, and the magnetic resonance image shows the singers vocal tract change shape to try to enhance different formants.

One can approximate the effect of the vocal tract on the amplitude of the harmonics by total amplitude of a harmonic = (source) x (filter function). This is illustrated in the figure on the following page. Note this figure is a qualitative representation of the physical concept.

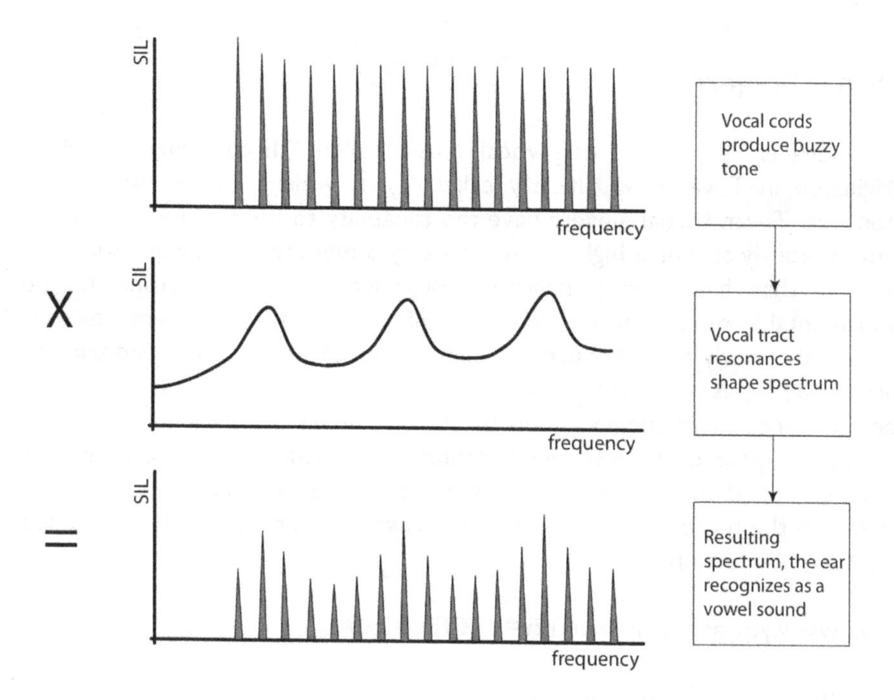

Singer's Formant

Professional male singers have an extra formant slightly higher in frequency than the third formant. This is done by lowering the larynx and widening the pharynx (cavity above the larynx). This high frequency peak allows the voice to be heard over an orchestra and is associated with a voice that projects well. This frequency (around 2600Hz) is above where the energy spectrum peaks for an orchestra.

Raised Formants

The soprano singing voice has frequencies that are often higher than the first formant. So a trained singer will raise her first formant to be in resonance with the fundamental frequency being sung. This often makes the high note parts sung by an opera singer difficult to understand. This must be taken into account in the composition by having the words repeated elsewhere in the performance, or by making the content of the words unimportant.

Throat Singers of Tuva

There is very interesting vocal music called "throat singing", which originated in Tuva (now officially called Tyva) which borders Russia and Mongolia. Tuvan throat singers have the capability to hold a low tone, while simultaneously singing a higher tone to carry a melody. The higher tone is a formant that has a very narrow resonance. The lower tone is the fundamental frequency of the vocal cord vibrations. The fundamental pitch is perceived as fairly loud through the phenomena of virtual pitch. The trick to throat singing is to produce a well-formed formant that has a narrow spectral range in frequency. Then the throat singer must be able to move the spectral peak of the frequency around to accentuate different harmonics in a controlled way. One can then make a musical melody using these pitches of the higher harmonics. Search "Tuvan throat singing" or check out this YouTube example:

http://www.youtube.com/watch?v=VTCJ5hedcVA

References and Resources

Bresch, E., and S. Narayanan "Real-time Magnetic Resonance Imaging Investigation of Resonance Tuning in Soprano Singing," J. Acoust. Soc. Am. **128** EL 335 (2010)

Rossing, T., F. Moore, and P. Wheeler, The Science of Sound, 3rd Edition, 2002

Sundberg, J., "Human Singing Voice," Encyclopedia of Acoustics, Vol. 4, p. 1687, M. Crocker Ed., (1997)

Chapter 12
Architectural Acoustics

Chapter 12 Outline

- Overview of Concert Hall Acoustics
- Qualities of Good Acoustics
- Acoustic Intimacy
- Source Width
- Shoebox Design
- Reverberation and Sabine's Formula
- Diffraction

We don't need to be working for a large architectural firm to have room acoustics impact us. When musicians routinely perform gigs in less than ideal venues like crowded rooms, small stages, or outdoors, they become intimately aware of the importance of acoustics. Even in your living room, or any room where you might be setting up a home theatre or preparing for a party, room acoustics play an important role. The large lecture classroom, church sanctuary, and the sports arena are other examples. I am sure you can think of many situations where you were bothered by poor room acoustics or poor sound amplification. In this chapter, we focus on concert hall acoustics, realizing that the concepts and ideas apply to many live acoustical situations. You may think of classical orchestral music being performed at a fine concert hall as a starting point to understand the basic concepts. Such venues are well studied and well tested, and they provide a case study for the other situations where you are concerned about room acoustics.

Macky Auditorium, University of Colorado, Boulder

Concert Hall Acoustics

Architectural acoustics has a bit of a dubious history, even though there are actually many more design successes than failures. Less than outstanding concert hall designs or re-designs come with a costly price tag, and their poor reviews make the news. Much of the science of room acoustics dates back to the early 1900's. Wallace Sabine, a Harvard Physics professor, took on the task of improving the acoustics in a poor-sounding lecture hall at the art museum at Harvard around 1895. Prior to that time, there was no science to it, and architects simply tried to imitate good sounding halls. Sabine later became famous with the excellent sounding Boston Symphony Hall which he helped design. In fact, Boston Symphony Hall is one of the best in the world.

However, the track record of architectural acoustics since then has been less stellar. The classic example is Philharmonic Hall at Lincoln Center for the Performing Arts that opened in Manhattan, in 1962 to mixed reviews, to put it politely. It was re-designed, re-opened, and renamed Avery Fisher Hall in 1976.

The older Carnegie Hall also in Manhattan continues to be considered an acoustically superior concert hall. Carnegie Hall was originally built in the 1890's and had major renovations in 1983 and 2003.

There are many interesting New York Times articles on concert hall acoustics relating to these two concert halls and others. If you are interested, it is worth searching the archives. For example, there is an excellent article in the "Science" section of the New York Times on November 9, 1997. There is also a short book titled "Deaf Architects and Blind Acousticians?" by Robert E. Apfel that is an excellent and concise resource if you desire further information on room acoustics [Apfel, 1998].

The big problem facing architectural acoustics in general (and not just for concert halls!) is trade-offs among:

1) visually pleasing design
2) use of space
3) acoustics

Just as important are constraints due to financing or the preferences of the people making the design decisions. Typically, if you are building or overhauling a concert hall, *compromises must be made between the parties:*

1) *The sponsor*, usually a committee or executive board, provides the architect and acoustical consultant with basic constraints (location, size, number of seats, type of performances, other uses, etc.)

2) *The architect* and acoustical consultant work out a rough conceptual plan and then meet with the sponsors. The sponsors then have input and make final decisions on finances.

There are *compromises that need to be made when planning* the architectural aesthetics and use of the planned concert hall:

1) Often, the architect is most interested in *making visual artistic or aesthetic impact.*

2) *Multiple uses of the hall:* theater, lecture, choral, orchestral, etc. Compromise between different uses makes optimal design difficult.

This give-and-take between the architect's interests, sponsor interests, and good acoustics is why the final result can become uncertain. If only the architects were completely blind and sponsors had extra piles of money to spend! However, visual cues and senses other than hearing obviously do play a role in our enjoyment of a concert. So, the visual (and all sensory) aesthetics is very important. We need a good architect. If visual cues are not consistent with what we hear, it just isn't quite as good. Otherwise, we could all just sit at home and use headphones to enjoy music! Well, we do that a lot don't we? Who has the money and the time for live concerts!

Qualities of Good Acoustics

There are seven important criteria to consider when designing a concert hall. I'm sure with a little thought you can come up with many of these common sense criteria listed below. For perspective, think of a large concert or lecture hall, say 500 seats, capable of musical performance, public lectures, and theater (for example, Macky Auditorium at the University of Colorado, Boulder).

Seven Criteria for Designing a Good Concert Hall

1) *Volume*

The sound must be loud enough for everyone to hear. In a concert hall, the volume of a tone held steady can build in the room as the sound reflects off the walls and ceilings. Concert halls should have a strong direct sound (from line of sight) and reflected sounds should build and not damp away too quickly. Volume is closely related to reverberation (discussed below).

2) The sound must be *well distributed* among the seating locations

All seating locations need to have decent acoustics. Or, those seats with poor acoustics will need to be steeply discounted.

3) *Clarity*, with no echoes and not too much reverberation

Clarity competes with reverberation and volume (volume builds as sound reflects off the walls). A room that is very clear is considered too "dead." A lecture hall that will be used for public speaking and presentations needs to emphasize clarity and sacrifice reverberation. Such a room will be acoustically "dead" relative to a concert hall. An opera hall requires more emphasis on clarity as well. A concert hall will have more reverberation and volume than an opera hall or a lecture hall. Distinct echoes are to be avoided. You avoid echoes by avoiding very large and reflective planar surfaces, such as walls, floors, and ceilings.

4) *Low background noise,* HVAC systems, traffic noise, etc.

A lot of expense and effort can go into reducing background noise. For concert halls, the optimum background noise is quite low and depends on frequency, but should be in the 20-30 dB range, or quieter than a library. Conventional HVAC "Heating, Ventilation and Air Conditioning" systems can generate noise well above these levels. In fact, HVAC systems can often be used as a source of white noise to mask other potentially annoying environmental noise found in commercial office buildings.

5) *Envelopment*

The sound appears to come from all directions. Think of the movie theater surround-sound effect. Envelopment is the result of reflections of sound off the walls and ceiling. The listener should feel "immersed" in the sound.

6) *Performer Satisfaction*

If the musicians are comfortable and enjoy performing, the audience benefits enormously. It is very important that the musicians can hear well and feel connected with the audience. Every effort should be made to make the acoustics on the stage of high quality.

7) *Good Reverberation*

> We will discuss reverberation in much more depth shortly, but for now, it is the tendency for sound to quickly build and slowly decay away due to multiple reflections. The decay should be slow, on the order of 1.5-2 seconds. The sound should linger. Good reverberant sound has a smooth decay without echoes. Flutter echoes are a common design flaw. These are pulsating "wooshes" of sound caused by reflections off large flat surfaces or walls, spaced far apart.

These seven criteria are simply good acoustical common sense. We can design for them knowing the absorption properties of building materials and how to estimate reverberation time. We will discuss how to do this shortly. Precisely, the reverberation time is the amount of time it takes for a steady sound to decay 60 dB when shut off.

There are two additional acoustical qualities that really separate great concert halls from the others. These two qualities truly "make or break" the acoustics:

1) *Acoustical Intimacy*
2) *Source Width*

Acoustical Intimacy

Listeners prefer acoustical intimacy.

Smaller rooms sound better. It is much easier to design a small room that sounds good than a large one. Why is this? Listeners simply prefer the quality of acoustical intimacy. But, how do we get this effect in a large venue? Well, you can actually judge room size even if you cannot see it. This physical size of a room can be perceived by hearing differences in the arrival time of *direct* and *reflected* sounds. Studies have found that the audience must hear a significant amount of reflected sound early for a good aesthetic. Specifically, the early reflections must be delayed no more than about 25 milliseconds. See the work of Leo Beranek, *Concert and Opera Halls: How They Sound* (1996), published by the Acoustical Society of America [Beranek 1996] for more discussion.

Source Width

Another very important quality that is also related to early reflections of sound is the quality of *source width*. In 1969, Stanford researchers P. Damask and V. Mellert, and later M, Schroeder and collaborators did studies using a dummy-head microphone. They concluded the following [Pierce 1992]:

Listeners like differences in the sound (or slight delays) between the left and right ear.

The term source width leads you to think we are talking about distances between the performers on stage, and to some extent this has an effect, but these dimensions or more-or-less fixed. When the paths of early reflections are different lengths we get differing delay times, and this is the more important contributor to the source width. We can design for this by choosing appropriate dimensions and materials.

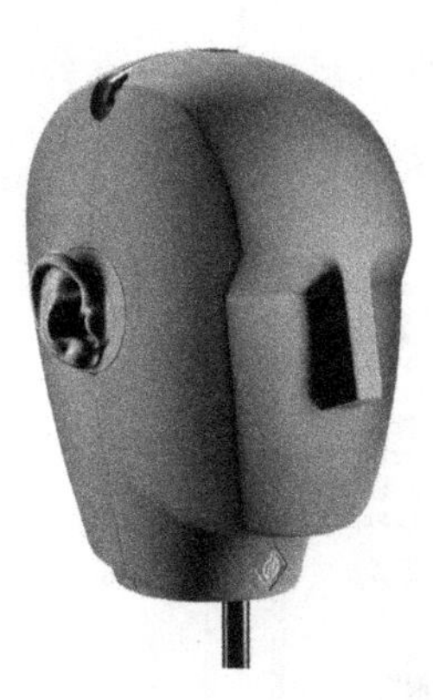

The Neumann KU 100 Dummy Head Microphone is available commercially for around $8,000. Courtesy of Georg Neumann GmbH.

These two criteria, acoustical intimacy and source width, naturally lead to the so-called shoebox concert hall design. Most of highly rated concert halls use this type of design.

Shoebox Design - Long, Tall, and Narrow

It is interesting that the most expensive seats (the center ones) are the ones that may suffer from a lack of source width in a shoebox shaped concert hall. This problem can easily be solved by adding asymmetries in reflecting surfaces near the front of the auditorium, causing early reflections to the two ears to be slightly different. One may see how to add such asymmetries by examining the figure below that shows only half of the concert hall. For example, one could design for asymmetries in the shape of the walls on the left and right side of the stage. The figure below also shows this delay time of the first reflection (1) arriving to the listener. It should be no longer than 25 milliseconds.

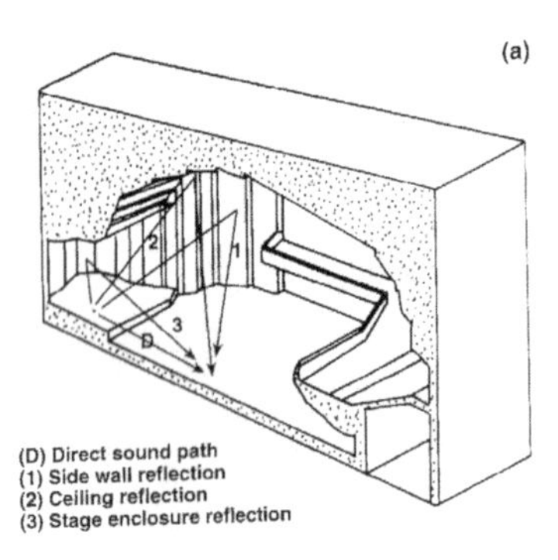

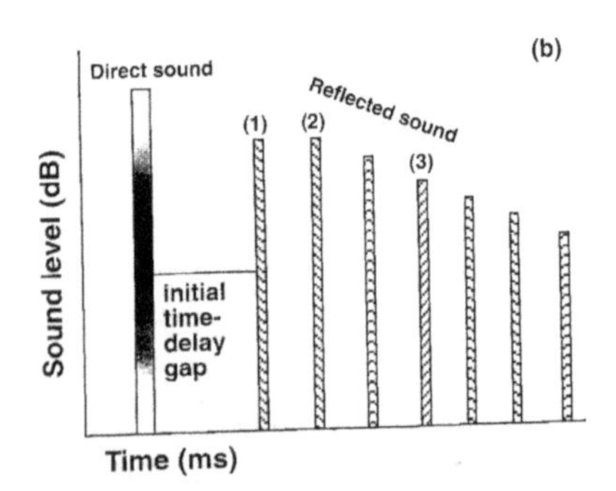

From R. E. Apfel (1998)

Reverberation

Reverberation is the way the sound builds in a room and later decays due to multiple reflections. Reverberation that decays away smoothly and is consistent with visual cues is important. However, good reverberation is often over-emphasized, and its importance over-rated. Why is this? Well, it turns out that Wallace Sabine (in the early 1900's) devised a mathematical way to describe reverberation and used it to design Symphony Hall in Boston, which is probably the best symphony concert hall in the United States. The reverberation time is relatively easy to characterize, but it turns out that some concert halls relying on Sabine's formula have been unimpressive. Examples with less than stellar ratings include the Philharmonic Hall (opened in 1962), the Royal Festival Hall in London (opened in 1951), and Toronto's O'Keefe Hall (opened in 1960), now called the Sony Centre for the Performing Arts. Of course, solely blaming Sabine's influence is a huge oversimplification, but we want to put the importance of the reverberation time in context. If you do any reading of introductory material you might walk away thinking it is the most important aspect in room acoustics, whereas it is, actually one of many. It just happens to be something we can easily calculate.

Wallace Sabine measured the reverberation time at the turn of the last century by firing a starter's pistol in a concert hall and measuring with a stopwatch how long it took the sound to decay. His definition is now standard and widely used, the reverberation time is:

Reverberation time, T_r - The time it takes for the sound level to decay 60 dB

Remember that the decibel or dB for short is a measure of the sound intensity level and is related to the psychological sensation of loudness. The reverberation time, T_r, can be measured for broadband noise, as would be done if one fired a starter's pistol in a concert hall. More sophisticated measurements involve using pure tones at different frequencies. The reverberation time is a function of frequency in such a case. It is important that the lower frequencies have a longer reverberation time. This is what gives warmth to the sound. A good reverberant sound is very important to the listener, it is something that listeners find pleasing. In fact, it is very common to add reverberation electronically. A second important aspect to reverberation is that it actually makes sounds louder. Reverberation helps build the sound intensity. The sound coming from a continuous source reflects off the walls, and the sound in the room builds over a short time. Then, when the sound source stops, the sound slowly decays away. This is shown in the figure below for Macky Auditorium at the University of Colorado, Boulder, CO. This sudden build up (after a few early reflections) and fairly smooth decay is what makes for a good reverberant sound. You do want to clearly hear the direct sound and early reflections so that there is intimacy and source width.

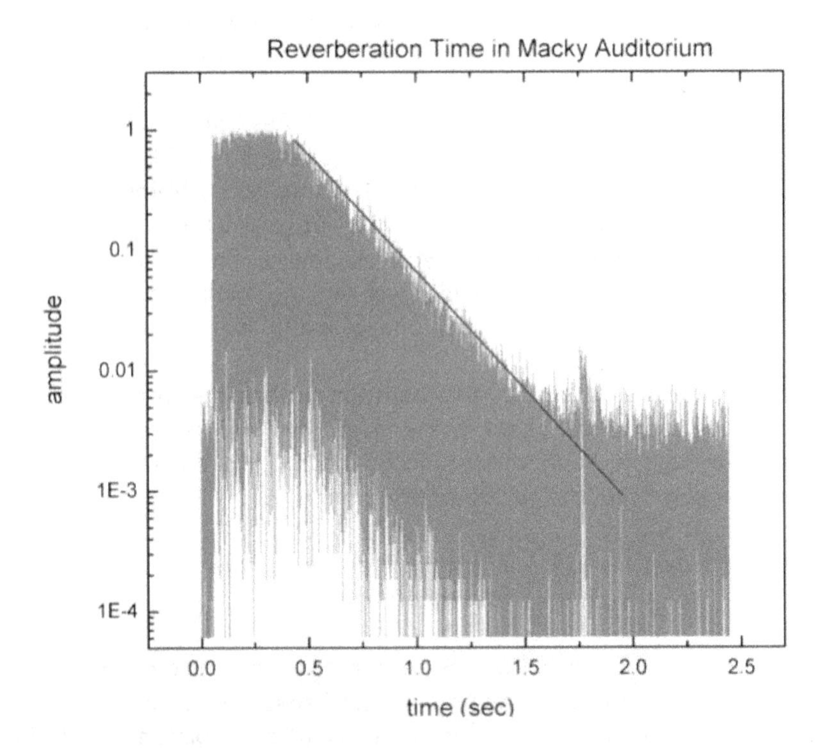

This is a figure of the reverberation time measured in Macky Auditorium, University of Colorado, Boulder (from J. Maclennan). Microphone voltage is proportional to pressure fluctuation amplitude. If the amplitude drops from 1 to 0.001, then the intensity drops from 1 to 1 millionth, which is a 60 dB drop. The result is a reverberation time for Macky Auditorium of about 1.5 seconds.

Reverberation times should be consistent with the visual cues. We know the size of the room by what our eyes see. A desirable reverberation time also depends on the type of music. Chamber music (a string quartet, for example) benefits from a shorter reverberation time than orchestral music. It is also important that the lower frequencies have a longer reverberation time, to add warmth to the sound.

Typical reverberation times for various size rooms

Practice room: $V = 27\ m^3$, $T_r = 0.6$ s (3m x 3m x 3m)

Rehearsal room: $V = 600\ m^3$, $T_r = 1.4$ s (6m x 10m x 10 m)

Large concert hall: $V = 20{,}000\ m^3$, $T_r = 2.2$ s (20m x 32m x 32m)

There can be problems associated with reverberation as well. These problems need to be avoided:

Flutter echo: A common problem occurs when there is distinct path of reflection off two hard walls. One hears fast, successive echoes or "flutter echoes."

Two distinct decay times: This occurs when there are two distinct chambers, or two distinct paths of reflection, which decay on two fairly different time scales.

Finally, having some asymmetries and different size objects tend to have a diffusive effect and smooth out the reverberation. Most concert halls have a large pipe organ. The pipe organ is considered important in improving the concert hall acoustics even if it is seldom played. Ceiling reflectors of different sizes and scalloped walls can contribute to a warm and smooth reverberant sound.

Sabine's Formula

We now discuss how to calculate the reverberation time using Sabine's Formula which is given by:

Sabine's Formula: $T_r = 0.16\ (V/S_e)$
(in seconds)

where T_r is the reverberation time in seconds, V is the volume of the room and S_e is the surface area of absorbing material. This formula is fairly general, and we will discuss, after the derivation that follows, how to determine S_e in this equation. You may skip the following page if you like.

This section (in a different font) is a little more advanced and may be skipped. If you have had calculus, you may follow the explanation of the origin of Sabine's Formula presented here. Let's assume that the walls in our room are completely absorbing. That is, any sound that hits the wall is absorbed. We can then exactly derive Sabine's Formula.

Let's say we play a steady tone in a closed room. We will assume that after a short time, the energy density ε, is uniform throughout the room.

$$\varepsilon = \text{sound energy density in Joules/m}^3$$

The power, P (Joules/sec), being dissipated by the absorbing walls will be the area of the walls times the intensity I (Watts/m^2) of the sound hitting the walls.

$$P = S\,I$$

The power lost to the walls must equal the time rate of change (Oh my gosh, calculus!) of the energy.

$$d/dt\,(\varepsilon\ V) = -P = -SI$$

V is constant and the intensity equals the energy density times the speed of sound divided by 4. We divide by 4 because sound is going out in all directions, left, right, forward, backward. But, what about sound going upward and downward? Actually, we do not divide by 6 because 4 gives a better answer (closer to what is actually measured). The physical reason is that sound is not really radiating uniformly in all directions due to floor and ceiling reflections, so the number is something less than 6.

$$d\varepsilon/dt\ V = -A\ v\ \varepsilon/4$$

$$d\varepsilon/dt = -Sv/(4V)\ \varepsilon$$

$$\varepsilon = \varepsilon_0 \exp\,(-Sv/(4V)\,t)$$

$$\ln\,(\varepsilon/\varepsilon 0) = -\ S\,v\,/\,(4\ V)\ T$$

The definition of reverberation time T_r, is a drop in the sound level of 60 dB, so $\varepsilon/\varepsilon 0 = 10^{-6}$.

$$T_r = -\ln\,(10^{-6})\ x\ 4\,/\,344\ V\,/\,S$$

$$T_r = 13.8\ x\ 4\,/\,344\ V/S$$

And Sabine's Formula for reverberation time follows:

$$T_r = 0.161\ V\,/\,S \text{ in seconds}$$

Absorption Coefficients

In reality, surfaces do not purely absorb or purely reflect. Hard smooth surfaces, such as cement or cinder block walls, are very reflective. Conventional drywall is reflective at higher frequencies, but can be surprisingly absorbing at lower frequencies depending on the supporting structural wall. *The absorption coefficient α, is defined as the amount of power absorbed divided by the amount of incident power on a given surface.*

The way we estimate the effective absorbing area S_e, in Sabine's Formula is by weighting absorbing areas by their absorption coefficient.

$$S_e = \alpha_1 \times S_1 + \alpha_2 \times S_2 + \alpha_3 \times S_3 + \ldots$$

S_i = surface area of each surface

α_i = absorption coefficient of the surface

S_e is the total "fully-absorbing" surface area of the room. Each surface is weighted by its absorption coefficient. You add up all surfaces. Each surface contributes its own surface area times its absorption coefficient. These calculations can become quite involved. You have to take into account the windows, doors, hallways, audience, empty seating, etc. Every surface has a different absorption coefficient. The absorption coefficient, α, is the fraction of energy lost each time the sound wave reflects off the surface. The values of α range from 0 to 1. A value of 0.01 would be highly reflective, 0.3 is fairly absorbing. Finally, to do reverberation time calculations, one needs a table of measured absorption coefficients, as shown below. These tables are common and can be found on the web. Note, that to use such a table, one must specify a frequency because reverberation time is dependent on frequency. If you are actually analyzing a particular room you should calculate the reverberation time for low, middle, and high frequency ranges.

Absorption Coefficients for Various Objects at Three Frequencies

Surface	Frequency		
	100 Hz	1000 Hz	4000 Hz
Drywall	0.3	0.04	0.1
Painted Concrete	0.1	0.07	0.1
Carpet	0.1	0.5	0.7
Window	0.3	0.1	0.04
Draperies	0.07	0.7	0.6
Upholstered Seating	0.2	0.7	0.6
Upholstered Seating with Humans	0.4	0.9	0.9

Data obtained from Table 15.1, p. 331, D. Hall (2002). Data presented here is simply to get a "feeling for actual numbers," see R. Apfel (1998) for more information.

Sabine's Formula is approximate. However, it works reasonably well if the average absorption coefficient is less than 0.15. That is, S_e / S_{total} < 0.15 where S_{total} is the total surface area of the room. It also gives a way to compare various options when choosing size, shape, and materials. There is also computer software that does "ray tracing." That is, the computer simulates the motion of the wave fronts and how they reflect off the surfaces and how they are absorbed (for example, CATT-Acoustic). These programs also have "diffusion." That is, a way to mimic how sound diffuses by scattering off of complex objects. These so called, "ray tracing" techniques are the same techniques used to simulate how light reflects off objects for three-dimensional visualization and animation. Another common practice is to make a scale model of the concert hall and test the model with higher frequency sound sources so that the wavelengths are in proportion to the size of the model. This was not done for Philharmonic Hall. What follows is a simple example problem involving calculating the reverberation time.

Example:

Calculate the reverberation time of a 20 m x 30 m concert hall with a ceiling height of 10 m. Assume the absorption coefficients for the walls, floor and ceiling are 0.1, 0.2, and 0.3 respectively.

Solution:

$T_r = 0.16\ V/S_e$

(Simply drop the units for Sabine's formula problems. It is best just to use numbers in metric units. If you write out the units, it gets messy when calculating S_e*)*

$V = 20 \times 30 \times 10 = 6000$

$S_e = 2 \times 0.1 \times 20 \times 10 + 2 \times 0.1 \times 30 \times 10 + 0.2 \times 20 \times 30$
$+ 0.3 \times 20 \times 30$

$S_e = 40 + 60 + 120 + 180 = 400$

$T_r = 0.16 \times 6000 / 400 = 2.4$ sec

Diffraction

Diffraction describes the way waves bend around corners in a diffuse way. Sound waves diffract. You can hear sounds around obstacles or through openings when you cannot see the source. Ah, but what about reflections? Don't they do the same thing? Yes, if I am in a room with the door open, I can hear sounds from the hallway without there being a "line-of-sight" to the source. Part of the reason is through multiple reflections. This is important to consider when designing a concert hall. But, even without multiple reflections, sound does bend around corners due to diffraction. For example, if you are standing behind a huge cement sound barrier, you can still hear lower frequency traffic noise coming over the barrier. Some vibrations are transmitted through solid material as well. Though light is an electromagnetic wave, its wavelength is very small, so it exhibits little diffraction in our day-to-day visual world. Very short wavelength sound waves behave like light in this way and cast shadows. Long wavelength sound waves diffract a lot. This is, in part, why low frequencies tend to be what you hear when someone is blasting a stereo in a car or a room. It also happens to be why the placement of the bass bin (bass speaker) is not important for sound re-enforcement or home theater design.

Low frequency vibrations tend to be transmitted more readily through structures (the walls). Why this is the case is complicated, but to absorb sound you need to convert the sound energy into heat. Absorption is enhanced with soft materials and holes and cavities on the scale of the wavelength of the sound. The transmission of vibrations through structures is not diffraction, but it is an additional effect worth keeping in mind.

We can make ballpark estimates of when sound (or any wave for that matter) will cast a shadow or diffract (or bend) around an object.

If the wavelength is smaller than the object, then there will be a shadow region.

Compare the size of the object L to the wavelength of the sound wave λ, where:

$\lambda = (344 \text{ m/s}) / f$

If $L > \lambda$, then little diffraction

If $L < \lambda$, then significant diffraction

For example, if we have a pillar 0.5 meters in diameter we would expect significant diffraction for sound waves with wavelengths larger than L. We can figure out the frequency range of diffraction by realizing there is a transition when $\lambda = L = 0.5$ m. We can calculate the frequency at this transition point:

$344 \text{ m/s} / (0.5 \text{ m}) = 688 \text{ Hz}$

We then expect significant diffraction for frequencies below 688 Hz.

The pillar would cast a shadow for frequencies above approximately 700 Hz. This would not be good acoustically (not to mention visually). However, even if the pillar were to the side of the listener, it would block early reflections and cause acoustical problems.

Just to calibrate our intuition, let's calculate wavelengths at a few frequencies:

65.4 Hz :	C2	: 5.3 m
130.5 Hz:	C3	: 2.3 m
261.6 Hz:	C4	: 1.3 m
523.25 Hz:	C5	: 0.66 m
1000 Hz:		: 0.34 m
5000 Hz:		: 7 cm
10,000 Hz:		: 3 cm

It is not too surprising we cannot hear above 20 kHz since at these high frequencies the wavelengths become smaller than the size of the ear.

Diffraction does not only apply to obstructions, but also to openings. For example, open windows or doors will allow directional sound to pass through the opening depending on the frequency range. Diffraction is very important for speaker design. If the diameter of the speaker cone is small compared to the wavelength of the sound, then sound will emanate in all directions. On the other hand, high frequencies with short wavelengths compared to the diameter of the speaker cone, will tend to be very directional. Bass bins or woofers can be placed most anywhere due to diffraction. Tweeters and midrange speakers need to be positioned very carefully due to their directionality.

This is not only true for speakers, but also true for musical instruments. If the diameter of the bell on a trumpet (or any horn) is larger than the wavelength of the sound produced, the sound will be very directional. The size of your mouth determines which frequencies are directional. Whispers are more directional due to their higher frequency content. If you play an instrument you should estimate at what frequency the sound should become more directional because it is something to consider in live performance situations.

Remember, most sounds have overtones and diffraction (the effect of waves to bend around corners) will apply to each frequency component and **not** just the fundamental frequency, which is often quite low (a few hundred Hz more-or-less).

References and Resources

Apfel, R., Deaf Architects and Blind Acousticians? A Guide to the Principles of Sound Design, 1998

Beranek, L., Concert and Opera Halls: How They Sound, 1996

Hall, D., Musical Acoustics, 3rd Edition, 2002

Pierce, J., The Science of Musical Sound, 1992

Chapter 13
Musical Illusion

Chapter 13 Outline

- Virtual Pitch
- Ascending and Descending Scale Illusions
- Categorical Perception
- Deutsch's Audio Illusions
- Deutsch's Octave Illusion
- Deutsch's Scale Illusion

In this last chapter, we will touch on more psychological or cognitive perceptual illusions relating to musical acoustics. These musical illusions are analogous to optical illusions that you may be more familiar with. For example, M.C. Escher used illusion in his artistic drawings quite extensively, see Escher's "Ascending and Descending" (1960), shown on the next page.

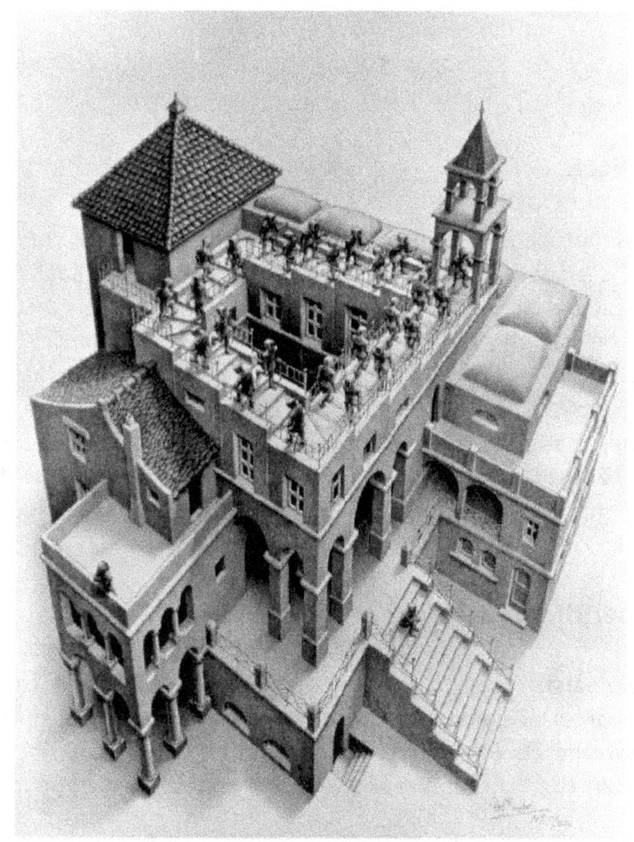

Escher's "Ascending and Descending" (1960)

Also see: http://www.mcescher.com/ for more on his work. Like Escher, you as a musician may integrate musical illusion into your creative work. One can create an illusion of an ever ascending or descending scale quite easily. We will explain this phenomenon shortly, but first, for an example of what these scales sound like, listen to YouTube:

http://www.youtube.com/watch?v=ev9hrqkhWsM

Virtual Pitch

The ever descending and ascending scale relies on the concept of virtual pitch. This is a subtle and remarkable perceptual effect. Let's briefly review what we learned about virtual pitch in Chapter 6. The concept of virtual pitch can be simply stated as *"the ear fills in the missing fundamental."* Mathematically, the ear perceives the pitch to be the greatest common factor of the overtones. Let's examine a concrete example. If you play one steady tone, made up by adding three pure tones with frequencies 800, 1000 and 1200 Hz, you would perceive a pitch of 200 Hz! The greatest common factor of these three frequencies is 200 Hz. It is quite remarkable that the ear does this.

Ever Ascending and Descending Scales

The Shepard's Scale is a famous illusion where one can create the perception of an ever-ascending (or descending) scale. Roger N. Shepard is a well-known psychologist in the area of cognitive science. Please return to (and listen to) the YouTube example of a Shepard's Scale above.

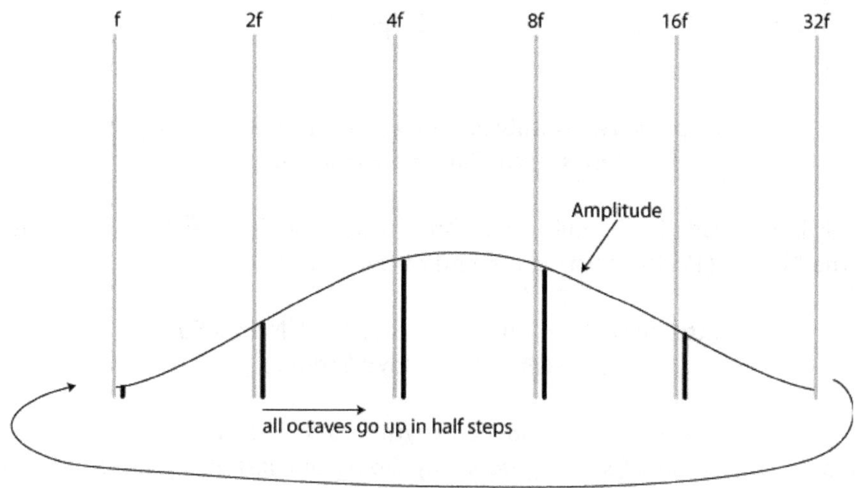

5 notes spaced an octave apart are shifted upward in half-steps. After twelve half-steps, the frequencies of all the notes are back exactly where they started. Note that the x-axis is logarithmic in frequency.

This mechanism works by continually shifting upward the frequencies of a few octaves, first with increasing, then decreasing intensity. The lower pitches are faded in with low amplitude, and the high pitch tones fade out with low amplitude. In the Shepard's Scale, the frequency of the harmonics are increased in half-steps. In the descending scales used by the electronic musician and composer Jean-Claude Risset, the partials are decreased continuously. When the partials move right to left (descending) continuously, it produces an eerie ever-decreasing pitch. This effect is found in:

Jean-Claude Risset, "Sud, Section III: "Afternoon, Evening" (1987)

There are many examples where this effect is used in performing arts to produce the nightmarish effect of a bomb dropping from an airplane. Jean-Claude Risset is famous for his use of illusion in music and some examples can be found on YouTube:

J-C. Risset - "Little Boy" (1968): www.youtube.com/watch?v=UKdJJvZIxPg

The Shepard-Risset effect is used in the movie "The Dark Knight" to give the effect of an ever-accelerating "Batpod."

See:

http://www.youtube.com/watch?v=FPaTv98wUsE
(or search Dark Knight Batpod)

It is also used in the quite long Pink Floyd song "echoes" on "Dark Side of the Moon" (1970), towards the end of the song. See:

http://www.youtube.com/watch?v=-KMpZaEF6g0
(or search Pink Floyd Echoes)

A similar effect can be produced rhythmically. Below is an example that gives the sensation of slowing down by fading out faster rhythmic patterns with slower ones. One could create a variety of rhythmic patterns that produce a "speeding up" or "slowing down" sensation without actually changing the meter or beats per minute.

In the figure above, the three rhythms are played together and produce a sensation of slowing down.

Categorical Perception

Our senses (sight, hearing, touch, smell, taste) produce an overwhelming amount of information and the brain quite efficiently processes this information by making a gross simplification of the situation. Though most stimuli vary in a gradual and continuous way, our brains tend to organize what we sense into discrete categories. This is really the only reasonable thing the brain can do. Suppose you had twenty thousand balls in your room with all shades of colors from the visible spectrum.

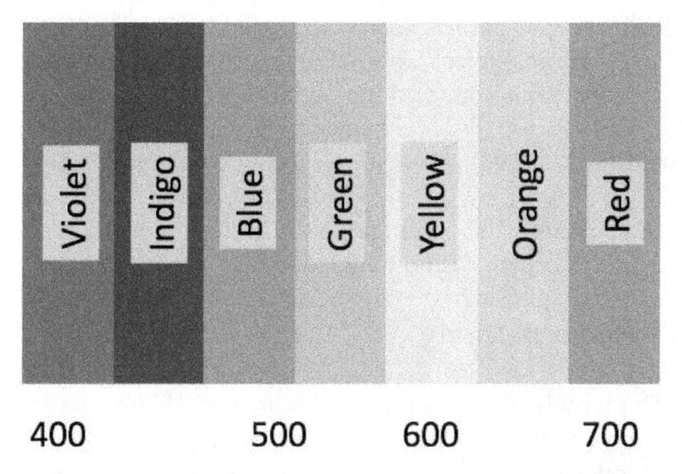

You decide to clean up and begin putting different color balls in different boxes. You wouldn't want to use too many boxes, otherwise, you might as well leave the balls out in the open and simply place them in order of color. You begin to make some decisions as to what goes in the blue box, the green box, the yellow box, and the red box. You might have a few more boxes, to place balls with "in between" shades of colors. The task of organization is fairly straightforward. This is the process called categorical perception and something your brain is doing all the time. The subtlety comes with the borderline cases. Do you put an orange ball in the yellow or red shoebox, for example? Maybe, you create another box for balls that are orange.

A good example of categorical perception is the tendency to categorize sounds into either/or categories: for example the "b" and "p" sound, or "g" and "k" sound. What if a sound falls somewhere between "b" and "p?" In fact, this is an important issue in speech and hearing science because someone with hearing loss cannot necessarily discriminate these different consonant sounds, yet they will tend to categorize them anyway (and often incorrectly). Humans tend to categorize objects (or sounds) as living or inanimate. We may categorize a sound as natural or electronic. This idea can be generalized to digital sound, as well as visual art. In a classic example, J-C. Risset's "Sud", which may be considered either electronic music or what is usually called "sound sculpture" or sound art. In his piece "Sud," Risset gradually morphs natural sounds into electronic sounds and produces an unusual psychological effect where the listener hears natural sounds one moment, then electronic sounds the next moment. Listen to the example:

J-C. Risset - "Sud" (1985): www.youtube.com/watch?v=RDsUypwnCQA

(You may simply search "Jean-Claude Risset Sud" to find this piece if the link is no longer active.)

Deutsch's Audio Illusions

Diana Deutsch is a psychologist who in the 1970's investigated (and discovered) a variety of auditory and musical illusions. Deutsch and her colleagues produced a CD that provides examples of some of these illusions. Deutsch is a professor of psychology at the University of California, San Diego. More recently, Deutsch has done ground-breaking research showing that people who speak tonal languages (Mandarin and Vietnamese) are much more likely to have perfect (or absolute) pitch than those who speak other languages. "Perfect pitch" is the ability to determine exact pitch of a sound, not just relative differences in pitch between two sounds.

Deutsch's Octave Illusion

Let us begin with the octave illusion. Octave intervals are played in the left and right ears. The right ear is played the high note then the low note ("high-low"). The left ear is played the low note then the high note ("low-high"). It is curious that right-handers perceive hearing only the high note in the right ear and the low note in the left ear. You may listen to this on Tracks 2-4 of Deutsch's CD, "Audio Illusions." Audio samples can be found at:

http://philomel.com/musical_illusions/

This illusion relies on a binaural effect and is best observed with ear buds or headphones. What happens? Typically, a right-hander will hear the high note in the right ear, and a left-hander will hear the high note in the left ear. What is odd is that this is true regardless of what ear the high note is played in. Put the headphones on backwards (switch the right and the left signal). You will still hear the high note in the right ear if you are a right-hander (in most cases).

The octave illusion can be embellished to produce unusual musical effects. The high-low illusion is very similar to the octave illusion, except now the pitches are sung. The high note is voiced with the word "high" and the low note is voiced with the word "low." The interval is still one octave. The same illusion is observed as for the octave illusion, where the high note is localized to the right-hander's right ear (usually). However, another illusion occurs as well. The words themselves seem to transform into different words as the pattern continues. Also, different sounds are heard depending on the balance between the left and right channels and the time delay between the left and right channel. So, as you move around in a room with loud speakers you will hear different things. People have reported hearing "blow pie," "high high," "pie pie," "buy loan," "no no," or "boat man" after listening for a while. Also, through stereo loud-speakers, the effect changes depending on your listening location. The octave illusion in this form may be used artistically.

Octave Illusion, from Figure 1 of the liner notes for D. Deutsch "Musical Illusions and Paradoxes" CD, http://philomel.com/musical_illusions/ Tracks 2-4

Deutsch's Scale Illusion

Deutsch found another interesting musical illusion, called the scale illusion where the brain synthesizes a major scale from a less regular pattern of notes presented to the right and left ear. This illusion is constructed from ascending and descending major scales with notes switching from left to right channel, as shown in the figure below. The actual patterns played to the left and right channels are shown in the first two lines.

Illustration of the Scale Illusion, From Figure 3 of the liner notes for D. Deutsch "Musical Illusions and Paradoxes" CD, http://philomel.com/musical_illusions/ Tracks 7-8

Listen to this pattern with headphones, and your sensory system takes a sound that is musically quite random (the top two lines) and makes sense out of it (bottom two lines). Your ear and brain are making musical order out of musical chaos! Reverse the headphones. Do you hear the same pattern? The bottom two lines shows what people typically hear, though other musical patterns may also be perceived. It is also interesting to experiment with listening to the scale illusion tracks using loudspeakers (without headphones) and adjusting the balance to see how this effect could be used artistically. The scale illusion clearly shows how two complex patterns are merged into two much more simple patterns by your brain. Like with the octave illusion, right-handers tend to hear the higher pattern in their right ear, whereas for left-handers there are less definitive results (regarding which pattern is heard in which ear).

This effect of making musical order out of chaos is more general than this one example and can be important musically. Suppose two instruments are playing parts that seem to jump around in an incoherent way. The listener may make sense out of the combined sound. This, of course, requires that a regular musical pattern exist within the series of notes of the two combined parts. The ear can, in many situations, find ordered musical patterns, like in the scale illusion.

We end this chapter by briefly discussing Deutsch's glissando illusion. This illusion is made up of an oboe sound that has a fixed pitch combined with a sine wave whose pitch glides up and down. Both sounds are alternated between the left and right channel. Listen to Tracks 12-13 on D. Deutsch "Musical Illusions and Paradoxes" CD using headphones. See:

http://philomel.com/musical_illusions/

What do you hear? Many people hear the oboe sound alternating from left to right, whereas the gliding tone seems to be well connected or balanced between the two ears. Listen to only one of the channels, using either the balance control on your stereo, or by just listening through one earphone. Notice that the sensation is quite different when listening to one channel. Many of these effects are especially important in electronic music where there is a lot of control over the sounds that go to the left and right channel.

In this chapter on musical illusion, we have gotten quite far from the physical nature of musical sound and have begun to explore the world of perception and cognitive psychology. We simply touch on some examples because they are interesting and may be used artistically. We also demonstrate that sometimes hearing and perception can be quite peculiar.

References and Resources

Further information on D. Deutsch's research -- http://deutsch.ucsd.edu

Deutsch, D., "An Auditory Illusion", *Nature* **251**, 307 (1974)

Diana Deutsch's Audio Illusions CD, Philomel Records, 1995, http://philomel.com/musical_illusions/

Further information on M.C. Escher's work -- http://www.mcescher.com/

Pierce, J., The Science of Musical Sound, 1992

Shepard, R., "Circularity in Judgments of Relative Pitch," Journal of the Acoustical Society of America **36,** 2346 (1964)

Index

A

B

C

D

R

S

T

CPSIA information can be obtained
at www.ICGtesting.com
Printed in the USA
BVHW040207260820
587333BV00016B/700

9 781482 566338